Design of Optimal Feedback for Structural Control

Ido Halperin
Department of Civil Engineering
Ariel University, Israel

Grigory Agranovich
Department of Electrical and Electronic Engineering
Ariel University, Israel

Yuri Ribakov
Department of Civil Engineering
Ariel University, Israel

CRC Press
Taylor & Francis Group
Boca Raton London New York

CRC Press is an imprint of the
Taylor & Francis Group, an **informa** business

A SCIENCE PUBLISHERS BOOK

First edition published 2021
by CRC Press
6000 Broken Sound Parkway NW, Suite 300, Boca Raton, FL 33487-2742

and by CRC Press
2 Park Square, Milton Park, Abingdon, Oxon, OX14 4RN

© 2021 Taylor & Francis Group, LLC

CRC Press is an imprint of Taylor & Francis Group, LLC

Library of Congress Cataloging-in-Publication Data

Names: Halperin, Ido, 1979- author. | Agranovich, Grigory, 1948- author. |
    Ribakov, Yuri, author.
Title: Design of optimal feedback for structural control / Ido Halperin,
    Grigory Agranovich, Yuri Ribakov.
Description: First edition. | Boca Raton : CRC Press : Taylor & Francis
    Group, 2021. | Includes bibliographical references and index.
Identifiers: LCCN 2020046707 | ISBN 9780367354121 (hardcover)
Subjects: LCSH: Structural control (Engineering) | Structural optimization.
    | Feedback control systems. | Control theory.
Classification: LCC TA654.9 .H35 2021 | DDC 624.1/71--dc23
LC record available at https://lccn.loc.gov/2020046707

ISBN: 9780367354121 (hbk)
ISBN: 9780367767006 (pbk)
ISBN: 9780429346330 (ebk)

Typeset in Times New Roman
by Radiant Productions

*In the memory of Prof. Jacob Gluck and
Prof. Vadim F. Krotov*

Preface

Structural control is a promising approach aimed at the suppression of unwanted dynamic phenomena in structures. It proposes the use of approaches and tools from control theory for analysis and manipulation of structures' dynamic behavior, with emphasis on suppression of seismic and wind responses. This book addresses problems in optimal structural control. It gathers some of our recent contributions to methods and techniques for solving optimal control problems that rise in structural control problems. Namely, it deals with solving optimal control design problems that are related to passive and semi-active controlled structures. The formulated problems consider constraints and excitations that are common in structural control. Optimal control theory is used in order to solve these structural control design problems in a rigorous manner, and based on a firm theoretical background. This monograph begins with a discussion on models that are commonly used for civil structures and control actuators. Modern theoretical concepts, such as dissipativity and passivity of dynamic systems, are discussed in the context of the addressed problems. Optimal control theory and suitable successive methods are over-viewed. Novel solutions for optimal passive and semi-active control design problems are derived, based on firm theoretical foundations. These results are verified by numerical simulations of typical civil structures subjected to different types of dynamic excitations.

The books is suitable for researchers and graduate students with background in civil engineering, structural control, control theory and basic knowledge in optimal control.

We would like to accentuate here two famous researchers whose contributions had significant influence on the results that are presented in this monograph—Prof. Jacob Gluck (1934–2001) and Prof. Vadim F. Krotov (1932–2015). Prof. Gluck was an Israel civil engineering researcher. He was an expert in seismic design as well as pioneer in structural computer-aided-design and structural control. In addition to being a famous researcher, he was a wonderful teacher and a pleasant person who devoted his life to advancing the field of civil engineering in Israel out of sincere concern for the lives and safety of people. Prof. Krotov was a honored Russian researcher, whose work focused on theory of nonlinear controlled dynamic systems and variational analysis in physics. His ideas have granted two main contributions to the theory of optimal control: a new theoretical approach to sufficient optimality conditions, and an iterative computational technique for deriving a sequence of monotonically improving control laws. This monograph is inspired by Prof. Gluck's vision and Prof. Krotov's bright ideas.

Contents

List of Symbols

$(v_i)_{i=1}^N$ A vector of order N with elements v_i

$\mathbf{v}$ A column vector of $(v_i)_{i=1}^N$

$\mathbf{V}$ A matrix

$(\mathbf{V})_{ij}$ Element ij of matrix $\mathbf{V}$

$\mathbf{diag}(\mathbf{v})$ A diagonal matrix with the vector $\mathbf{v}$ in its main diagonal

sign   The customary sign function $\mathbb{R} \rightarrow \{-1,0,1\}$

$\mathbf{A}$ State equation dynamic matrix

$\mathbf{B}$ State equation control input matrix

$\mathbf{C}$ Output equation state observation matrix

$\mathbf{C}_d$ Damping matrix

$\mathbf{D}$ Output equation input observation matrix

$e^{\mathbf{A}}$ The exponential matrix of $\mathbf{A}$

$\mathbf{\Psi}$ DOF's control forces input matrix

G Shear modulus

$\mathbf{\Phi}$ State transition matrix

$\mathbf{K}$ Stiffness matrix

$\mathbf{M}$ Mass matrix

n The plant order

n_u The number of control inputs

n_w The number of control forces

n_z The number of dynamic DOFs

$\mathbf{g}$ State ground acceleration input vector, state equation excitation input trajectory

$\mathbf{Q}$ State cost

$\mathbf{R}, r_i$ Control input cost

γ Shear strain

$\boldsymbol{\gamma}$ DOFs ground acceleration input vector

t Time variable

$\mathbf{f}$ State derivatives vector function

$\mathbf{f}^c$ Constraints equation

$\mathbf{h}$ Output equation vector function

H Hamiltonian function

$\mathbf{p}$ Costate/Lagrange multipliers vector function

J Performance index

l_f, l Terminal and integrand cost functionals

τ	Shear stress	$\mathbf{f_x}$	The jacobian matrix of trajectory $\mathbf{f}$ with relation to $\mathbf{x}$
μ	KKT multiplier		
$\mathbf{x}$	State trajectory	$\mathcal{U}$	The set of admissible control trajectories
$\mathbf{y}$	Output trajectory		
$\mathbf{u}$	control input trajectory	$\mathcal{H}$	The set of admissible state trajectories
$\mathbf{\hat{u}}$	Feedback form for $\mathbf{u}$		
$\mathbf{w}$	Control forces trajectory	q	Improving function
$\mathbf{z}$	DOF displacements trajectory	CBQR	Constrained bilinear quadratic regulator
$\mathbf{\dot{z}}, \mathbf{\ddot{z}}$	The first and second full derivative of trajectory $\mathbf{z}$	CBBR	Constrained bilinear biquadratic regulator
$z_d, \dot{z}_d$	MRD elongation, elongation rate and state vector	DOF	Dynamic degrees of freedom
		KKT	Karush-Kuhn-Tucker
$\ddot{z}_g$	Ground acceleration	LQ	Linear quadratic
$\mathbb{R}, \mathbb{R}^{n \times m}$	The set of real scalars, the set of real $n \times m$ matrices	LQR	Linear quadratic regulator
		LTI	Linear time invariant
$\mathbf{M}^{-1}$	The inverse of a matrix $\mathbf{M}$	MR	Magnetorheological
$\mathbf{M}^T$	The transpose of a matrix $\mathbf{M}$	MRD	Magnetorheological damper
e_g, e_{gg}	The gradient (row vector) and the hessian matrix of a function e	MRF	Moment resisting frame
		VD	Viscous dampers

Chapter 1

Introduction

Vibration control is a major problem in many mechanical systems. Continuous mild vibrations may interfere with the system's performance and/or cause a rapid weathering of physical elements. Intensive vibrations may even cause immediate damage to the system. A promising approach for dealing with such unwanted dynamic phenomena in civil structures, has attracted growing attention over the last decades. This upcoming field, known as structural control, suggests the use of approaches and tools from control theory for analyzing and/or improving structures dynamic behavior, with emphasis on suppressing seismic and wind responses of structures [89]. Its goal is to keep the structures strains, stresses, accelerations, displacements, etc., which are caused by a dynamic excitation, below a given bound [49].

Generally speaking, physical realization of structural control is done by applying forces to the vibrating structure in real time. The required forces are generated by mechanical devices, called *actuators*, which take commands (*control signals*) from a *controller*. Control implementation consists of two main stages. First, a law, which defines what are the suitable control signals at each moment, should be formulated. Next, this law is embedded into some electronic controller that calculates these signals in real-time and translates them into electric signals. These electric signals are sent to the actuators, which turn them into real forces that are applied to the structure. Various types of actuators, have been developed for that purpose. Each type of device is based on different physical phenomenon and possesses its advantages and limitations. This led to the definition of four main structural control classes: passive, active, semi-active, and hybrid [49].

In active control systems, the actuators need external energy source in order to generate control forces. The actuators are distributed in the structure and can be used to add or dissipate mechanical energy from it. Although this strategy is known to be highly effective, the need for a significant energy source makes active control vulnerable to power failure, which is very likely to occur during strong excitations, i.e., earthquakes. This disadvantage brings into question the effectiveness of active control for improving structural response to extreme loadings [49].

Passive control is a well known solution for structural vibration problems. An important feature of a passive control system is that it can generate the required control forces without any external power. Instead, the structural motion itself is utilized to generate the forces in dampers by forcing motion on the damper's anchors. Essentially, passive control has four main advantages: (1) it is usually relatively inexpensive, (2) it consumes no external energy, (3) it is inherently stable and (4) it works even during strong excitations [49]. However, a major drawback of passive methods is that they are unable to adapt to changes in structural properties, usage patterns and loading conditions. For example, structures that used passive base isolation in one region of Los Angeles and survived the 1994 Northridge earthquake, may have been damaged severely if they were located elsewhere in the region. Yet, its relative simplicity and reliability makes passive control a worthy alternative in many structural control problems.

Semi-active control strategy is sometimes referred to as a class of active control. Similar to an active one, a semi-active control system can alter its properties in real-time, however, it can do this by a very small amount of external power, compared to an active control system [49]. An important group of semi-active devices are semi-active *dampers*. The forces in such dampers can only resist the structural motion in the damper's anchors but the device's resistance properties can be changed by a very small amount of external power [49]. Semi-active systems provide an attractive alternative to active ones for structural vibration reduction. They offer the reliability of passive devices and maintain the versatility of fully active systems without the need for a large power sources. However, at the same time semi-active dampers set restrictions on the control forces, which significantly complicate the use of optimal control theory for controller design.

A hybrid control system is a combination of different control schemes. In many cases, the only essential difference between active and hybrid control is the amount of external power used for control implementation. A side benefit is that in case of power failure the system still provides a certain protection to the structure.

The present book deals with several optimal control design problems, which are related to passive and semi-active controlled structures, and with their solutions. First, the problems are formulated while taking into consideration constraints and excitations which are common in structural control. Next, optimal control theories are used in order to solve these problems rigorously. Chapter 2 analyzes models that are commonly used for civil structures and control actuators. The analysis is performed by modern theoretical concepts, such as systems dissipativity and passivity, and hints to novel models and approaches for optimal control solutions. Chapter 3 describes optimal control theories that are suitable to the addressed problems. Chapter 4 introduces successive methods that are used later for solving optimal control problems, related to control law design. Chapters 5 and 6 present new results that correspond to optimal passive and semi-active control of structures, respectively. Chapter 7 presents an approach for effectively placing dampers in structures with seismic excitation.

Chapter 2

Dynamic Models of Structures

This chapter presents an overview on models, which are commonly used for civil structures and control actuators, and that are relevant to the problems that will be addressed in the following chapters. Afterwards, fundamental terms in dissipative systems theory are discussed and interpreted within the context of structural control.

2.1 Plant Models

A plant model is essentially a mathematical abstraction of a physical system [80] that brings into account all the corresponding inputs, outputs and their dynamic relations. In control theory, it is common to use the continuous time, state space equations, for plant modelling. This model is composed of a set of first order differential equations:

$$\dot{\mathbf{x}}(t) = \mathbf{f}(t, \mathbf{x}(t), \mathbf{u}(t)); \quad \mathbf{x}(0), \forall t \in (0, t_f) \tag{2.1}$$

and a set of algebraic equations:

$$\mathbf{y}(t) = \mathbf{h}(t, \mathbf{x}(t), \mathbf{u}(t)) \tag{2.2}$$

where $t_f \in (0, \infty)$ and $(0, t_f)$ is a time interval; $\mathbf{x} : \mathbb{R} \to \mathbb{R}^n$ is a vector function, denoted as the *state* trajectory. It represents a set of fundamental processes that forms a foundation for the plant's dynamic response. $\mathbf{x}(t)$ is an abbreviation for $(x_1(t), x_2(t), \ldots, x_n(t))$; $\mathbf{u} : \mathbb{R} \to \mathbb{R}^{n_u}$ is a vector function that represents a set of control inputs signals. These signals are those that are being used for the manipulation of the plant's response. In this book it will be denoted also as *control* trajectory; $\mathbf{f} : \mathbb{R} \times \mathbb{R}^n \times \mathbb{R}^{n_u} \to \mathbb{R}^n$ is a mapping that defines the plant dynamic properties; $\mathbf{y} : \mathbb{R} \to \mathbb{R}^{n_y}$ is the *outputs* trajectory and $\mathbf{h} : \mathbb{R} \times \mathbb{R}^n \times \mathbb{R}^{n_u} \to \mathbb{R}^{n_y}$ is a mapping that defines the relations between the output trajectory and the state and input trajectories.

In many control problems a great significance is given to formulating $\mathbf{u}$ in a *feedback* form. A feedback is a mapping $\hat{\mathbf{u}} : \mathbb{R} \times \mathbb{R}^n \to \mathbb{R}^{n_u}$ such that:

$$\mathbf{u}(t) = \hat{\mathbf{u}}(t, \mathbf{x}(t)) \tag{2.3}$$

It is customary to refer the trajectories, generated by $\hat{\mathbf{u}}$, as *closed loop* control trajectories. Otherwise they are said to be *open loop* ones. As known, closed loop signals benefit improved disturbance rejection and reduced sensitivity to parameter variations and uncertainties [70]. On the other hand, its synthesis is more complex and much more challenging.

Any computer aided control design begins at the plant modeling stage, that is, the formulation of $\mathbf{f}$ and $\mathbf{h}$. Basically, there are two main approaches to construct a model. Physicists will be interested in models which reflect, clearly as possible, the underlying physical mechanisms of the observed phenomena and that are not falsified by the available experiments. This is an analytic approach, which develops the model from its physical principles [95].

When using this approach for structural systems, either Newton's 2nd law or Euler-Lagrange equations is commonly used to derive a set of differential equations by means of some Cartesian and/or generalized coordinates. It is customary to use the term *dynamics degrees of freedom* (DOFs) to refer coordinates that simply describe the geometric displacements and/or rotations at different locations throughout a structure. The equations of motion, which are derived in the DOFs of a model with lumped masses, linear damping, linear stiffness and a set of actuators, are given by the following 2nd order initial value problem [49, 90]:

$$\mathbf{M}\ddot{\mathbf{z}}(t) + \mathbf{C}_d\dot{\mathbf{z}}(t) + \mathbf{K}\mathbf{z}(t) = \mathbf{\Psi}\mathbf{w}(t); \quad \mathbf{z}(0), \dot{\mathbf{z}}(0), \forall t \in (0, t_f) \tag{2.4}$$

It is a *linear time invariant* (LTI) model, in which $\mathbf{M} > 0$, $\mathbf{C}_d \geq 0$ and $\mathbf{K} > 0$ are symmetric mass, damping and stiffness matrices, respectively. The relations $>$ and $\geq$ are the customary notations for positive definiteness and non-negativity of a matrix, respectively; $\mathbf{z} : \mathbb{R} \to \mathbb{R}^{n_z}$ is a vector function that describes the displacements in the DOFs; $\mathbf{w} : \mathbb{R} \to \mathbb{R}^{n_w}$ is a vector function of the control forces inputs that are applied to the plant and $\mathbf{\Psi} \in \mathbb{R}^{n_z \times n_w}$ is an input matrix that describes how $\mathbf{w}$ affects the structure. This model is frequently used to describe the dynamics of civil structures [90]. From physical viewpoint, $\mathbf{M}$ represents the inertial properties of the system; $\mathbf{K}$ represents its elasticity and $\mathbf{C}_d$ its energy dissipation properties. Although $\mathbf{C}_d$ is often considered classical, in this book it is considered general, i.e., the mode shapes are unnecessarily orthogonal [21].

For control problems, in which the control trajectories represent control forces (i.e., $\mathbf{u} \equiv \mathbf{w}$), the state space form of Eq. (2.4) takes the following form:

$$\dot{\mathbf{x}}(t) = \mathbf{A}\mathbf{x}(t) + \mathbf{B}\mathbf{u}(t); \quad \mathbf{x}(0), \forall t \in (0, t_f) \tag{2.5}$$

where

$$\mathbf{x}(t) = \begin{bmatrix} \mathbf{z}(t) \\ \dot{\mathbf{z}}(t) \end{bmatrix}; \quad \mathbf{A} \triangleq \begin{bmatrix} \mathbf{0} & \mathbf{I} \\ -\mathbf{M}^{-1}\mathbf{K} & -\mathbf{M}^{-1}\mathbf{C}_d \end{bmatrix} \in \mathbb{R}^{n \times n}$$

$$\mathbf{B} \triangleq \begin{bmatrix} \mathbf{0} \\ \mathbf{M}^{-1}\mathbf{\Psi} \end{bmatrix} \in \mathbb{R}^{n \times n_u}$$

$n = 2n_z$ and $n_u = n_w$. Usually the plants, addressed in structural control, are asymptotically stable in nature, which implies that $\mathbf{A}$ is Hurwitz.

For $\mathbf{y}$ that is linear with respect to $\mathbf{x}$ and $\mathbf{u}$, the corresponding algebraic output equation will be

$$\mathbf{y}(t) = \mathbf{C}\mathbf{x}(t) + \mathbf{D}\mathbf{u}(t) \tag{2.6}$$

where $\mathbf{C} \in \mathbb{R}^{n_y \times n}$ and $\mathbf{D} \in \mathbb{R}^{n_y \times n_u}$ are constructed according to the defined $\mathbf{y}$.

When the structure is subjected to seismic excitation, the plant state equation will be

$$\dot{\mathbf{x}}(t) = \mathbf{A}\mathbf{x}(t) + \mathbf{B}\mathbf{u}(t) + \mathbf{g}\ddot{z}_g(t); \quad \mathbf{g} \triangleq \begin{bmatrix} \mathbf{0} \\ -\boldsymbol{\gamma} \end{bmatrix}, \quad \mathbf{x}(0), \forall t \in (0, t_f) \tag{2.7}$$

where $\ddot{z}_g : \mathbb{R} \to \mathbb{R}$ is a function that describes the ground acceleration and $\boldsymbol{\gamma} \in \mathbb{R}^{n_z}$ is a vector that describes the direct influence of the ground acceleration on the DOFs.

For engineers, who are usually more interested in the engineering application rather than the model itself, the analytic framework is sometimes much too involved to be really useful. From a practical viewpoint, a model that accurately describes the system behavior, but lacks a clear physical meaning, is preferable than a less accurate one that emphasizes the physical laws governing the system. From this reason, engineers will typically use system identification techniques to build their models [95].

By the system identification approach, modeling of dynamic systems is done from experimental data. First, experiments are performed on a given system. Next, a certain parameterized model class for this system is predefined by the user. Finally, suitable numerical values are assigned to the parameters so it will fit as closely as possible to the recorded data.

There are quite a number of system identification methods that address the formulation of LTI models. Apparently, since it is known that most phenomena are intrinsically non-linear, the use of LTI models might be questionable. However, it is justified because:

■ Experience shows that vibrations in civil structures are well approximated by linear finite dimensional models and that sometimes, complex behavior can be taken into account by choosing a high enough plant order—n.

■ Controller design is much easier for linear finite dimensional models. As a matter of fact, this is the only class of models for which computer aided control design is actually tractable in full generality, and for which there is a complete rigorous theory available [95].

■ An effective control design reduces the seismic response such that the structural members behavior is mainly linear.

Yet, sometimes nonlinearities might still have noticeable contribution to the response of such approximated LTI plant. In order to cope with such nonlinearities in the control design and still remain with an LTI model, two options are available [95]. Either the system can be described as a time-varying one by using a recursive identification for model updating. This approach essentially corresponds to local linearization of the nonlinear system. Or, it is possible to use a robust control design. Observations show that mild nonlinearities do not matter as they can be incorporated in the control design by means of robustness for dynamic uncertainties.

Bilinear Models

For plants that can't be described accurately enough by linear models, it is inevitable to consider the formulation of a non-linear model.

Bilinear models are simple and at the same time effective nonlinear dynamic models. They can be used for approximation of nonlinear systems at a common fixed point and in certain practical modern control problems [100, 57]. Despite their nonlinearity, their properties are close to those of linear systems and therefore can be treated by several techniques and procedures from the linear systems' theory [14].

A general, continuous time, bilinear model has the form:

$$\dot{\mathbf{x}}(t) = \mathbf{A}(t)\mathbf{x}(t) + \mathbf{B}(t)\mathbf{u}(t) + \sum_{i=1}^{n_u} \mathbf{N}_i(t)\mathbf{x}(t)u_i(t); \quad \mathbf{x}(0), \forall t \in (0, t_f) \quad (2.8)$$

where $\mathbf{A} : \mathbb{R} \to \mathbb{R}^{n \times n}$, $\mathbf{B} : \mathbb{R} \to \mathbb{R}^{n \times n_u}$ are matrix functions and the bilinear element is defined by a set of matrix functions—$(\mathbf{N}_i : \mathbb{R} \to \mathbb{R}^{n \times n})$. When $\mathbf{A}$, $\mathbf{B}$, $(\mathbf{N}_i)$ are constant, the system is *time invariant*. When $\mathbf{B} = \mathbf{0}$ the system is said to be *homogeneous in state*.

The state-space equation, described in Eq. (2.8), highlights an important advantage of bilinear models, even though they are nonlinear, bilinear models are very "close" to linear ones. Additionally, bilinear models are sometimes referred to as *variable structure linear models*. From this viewpoint, these models are regarded as linear models with controlled parameters [14]. This allows to take advantage of known techniques and procedures from the theory of linear models for controlling bilinear models [14].

Optimal control of bilinear models have been addressed by several researchers. A bilinear quadratic optimal control problem is defined for a bilinear system and quadratic performance index [1]. No constraints were imposed on the control. An iterative scheme, producing a linear control law, was suggested. It was derived by Pontryagin's minimum principle and the successive approximation approach. Essentially, the method requires solving the differential Lyapunov equation at each iteration. A proof of convergence was given. The use of Adomian's decomposition method for solving a time-varying bilinear quadratic, optimal unconstrained control problem, was suggested [22]. According to this method, the bilinear model is represented by a convergent sequence of linear ones. Then the solution of bilinear quadratic problem is represented as the convergent sequence of LQR solutions. Optimal tracking problem was used to illustrate this method. An optimal control law for a bilinear system and a quadratic cost for tracking problem, was suggested [62]. The theoretical framework was constructed using Lagrange's multipliers and an iterative algorithm was proposed for its solution. The algorithm is based on redefined bilinear system variables. A proof of convergence was provided. The bilinear quadratic optimal unconstrained control problem is solved by construction of Hamiltonian equations, which leads to the common two-point boundary value problem for the state and costate. The computation of the initial costate is done successively by solving two first-order quasi-linear partial differential equations [26]. An optimal control of bilinear model with control signal bounds was treated [12]. A stochastic Hamilton-Jacobi-Bellman equation was used for formulating the optimal control by means of boundary value problem. The problem was not solved, but it was used as a theoretical basis for a corresponding clipped optimal control.

2.2 Viscous and Semi-active Dampers

Structural control realization is done by mechanical actuators that apply control forces to the structure. However, since each type of actuators has its special advantages, disadvantages and limitations, suitable restrictions should be taken into consideration already in the stage of problem definition. This dependency creates a strong relation between the type of actuators that will be used and the definition of the control problem. This section reviews several types of actuators and stress properties which affects the control problem definition.

A well known class of devices used for passive control of mechanical systems, is viscous fluid dampers (VD). These are pistons, which contain viscous fluid such that when the damper changes its length, a flow is generated through configured orifices. The viscosity of the flowing fluid creates force that resists the damper's motion. A schematic view of a VD is given in Fig. 2.1.

Installing the damper in the structure forces a geometric compatibility between the damper and the structure. As a result, the structure oscillation induces motion on the dampers and generates damping forces on

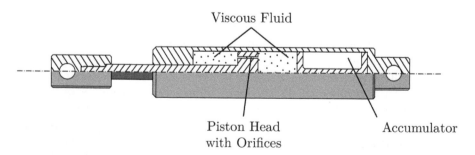

Figure 2.1: A schematic view of a viscous fluid damper.

the structure. In general, the tensile force generated by the damper is governed by:

$$w(\dot{z}_d(t)) = c_d |\dot{z}_d(t)|^\alpha \operatorname{sign}(\dot{z}_d(t)) \tag{2.9}$$

where c_d is the VD's *damping constant*, also known as *viscous gain*. Although these two terms are practically identical, their semantic is slightly different. The former describes how c_d affects the system's dynamic response whereas the latter is inspired by its interpretation when using input-output VD formulation; $\dot{z}_d$ is the damper's elongation rate which is imposed by the relative velocity of the damper's anchors; sign $: \mathbb{R} \to \{-1, 0, 1\}$ is the customary sign function and α is the velocity exponent. Usually, the values of α varies between 0.15 and 2.0. During the design stage, it is common to use $\alpha = 1.0$ such that Eq. (2.9) simplifies to [49]:

$$w(\dot{z}_d(t)) = c_d \dot{z}_d(t) \tag{2.10}$$

This linear form is more tractable and therefore more convenient from design viewpoint. After the desired c_d was evaluated, further design refinement can be done by using other values for α [11, 7].

Another famous class of energy dissipating devices is the semi-active damper. An essential feature of these dampers is that their properties (i.e., viscosity, friction, etc.) can be rapidly adjusted in real time by no more than a low power source. A type of semi-active device, which recently attracts the attention of many researchers, is the magnetorheological damper. This device takes advantage of non-newtonian fluids, which are characterized by rapid apparent viscosity changes due to a varying magnetic field. These fluids are called magnetorheological (MR) and allow adjusting the damper characteristics according to the varying needs of the system [9].

MR fluids typically consist of micron-sized ferrous particles dispersed in a working fluid. The change in the rheological behavior of the fluid results from an external magnetic field, which is induced on the suspended particles and polarizes them. The induced dipoles cause the particles to align "head to tail" in chains and form columnar structures, parallel to the applied field. These chain-like structures hinder the flow of the fluid and thereby increase its apparent viscosity. The pressure needed to yield these chain-like structures increases with the applied magnetic field, resulting in a field-dependent yield stress [37]. In the pre-yield phase, the MR fluid behaves as a viscoelastic solid and therefore, for small strains, the shear stress τ is given by [37]:

$$\tau(\gamma, B) = G(B)\gamma$$

where G is the shear modulus; B is the applied magnetic field and γ is the shear strain. It has been observed that G is a magnetic field dependent [37]. In the post-yield phase, the behavior is often represented as a Bingham plastic model having a variable yield strength and an apparent viscosity [99]:

$$\tau(\dot{\gamma}, B) = \tau_y(B) + \kappa(B)\dot{\gamma}$$

where τ_y is the yield stress; κ is the viscosity of the MR fluid and $\dot{\gamma}$ is the shear strain rate. τ_y and κ depends on B.

MR *dampers* (MRDs) are similar to VD but instead of conventional viscous fluid they contain an MR one. By applying a magnetic field to the MR fluid, which flows inside the damper, the fluid turns into a semi-solid material. This allows changing the physical dissipative characteristics of the damper in real time. MRDs have the advantage of low maintenance, low activation power demand and excellent performance [49]. An MRD is schematically described in Fig. 2.2.

The dependency of the MRD force on the input current was studied by many researchers [99]. To this end, different models and approaches were suggested, most of them are non-linear. Examples are the hyperbolic tangent function [34] and Bouc-Wen based models [82, 99]. In the latter, each MRD add another dynamic state to the system. This state is also experimentally justified by the typical hysteretic behavior of MR fluid-based devices. However, a simple approach for MRD modeling is to assume that the force follows Bingham's constitutive model. In this case, the tensile force in the MRD is governed by [99]:

$$w(\dot{z}_d(t), i_c(t)) = c_d \dot{z}_d(t) + w^y(i_c(t)) \operatorname{sign}(\dot{z}_d(t)) \qquad (2.11)$$

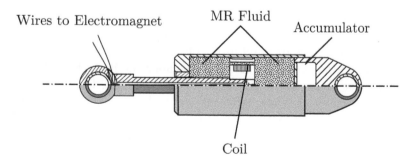

Figure 2.2: A scheme of an MR Damper.

where $i_c : \mathbb{R} \to \mathbb{R}$ is the MRD input current; $c_d \geq 0$ is the MRD viscous damping gain and $w^y : \mathbb{R} \to [0, \infty)$ is a controllable yield force. In this model it is usually assumed that c_d is constant while w^y may be varied according the control law [34].

A variable viscous damper [78] is a type of semi-active device, whose damping gain is controlled by real-time adjustment of the mounting angle between two viscous fluid dampers. Let $\theta : \mathbb{R} \to [\theta_{min}, \theta_{max}]$ be the time varying mounting angle of the viscous fluid dampers. The tensile force is

$$w(\theta(t), \dot{z}_d(t)) = 2\sin(\theta(t))^2 c_d \dot{z}_d(t)$$

Semi-active friction damper was suggested for structural control [36]. A scheme of the damper is given in Fig. 2.3. It consist of an external cylinder (1), two internal half-cylinders (2a and 2b) and a pneumatic vessel (3). The tensile force in this damper is:

$$w(\dot{z}_d(t), p(t)) = w^f(p(t)) \operatorname{sign}(\dot{z}_d(t))$$

where $w^f : \mathbb{R} \to [0, \infty)$ is the slip force and $p : \mathbb{R} \to \mathbb{R}$ is the applied pressure in the pneumatic vessel. The changes in p alter the level of the slip force and allows for the damping force to be adjusted.

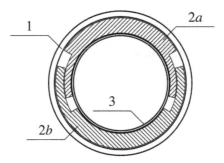

Figure 2.3: A scheme of a semi-active friction damper [36].

Additional types of semi-active devices, which can be found in the literature, are electro-rheological devices [79], electro-magnetic friction dampers [3], variable orifice dampers [89, 65], etc.

Both VDs and semi-active dampers are inherently dissipative control devices [51]. Here 'inherently dissipative' or 'physical dissipativness' means that physical considerations assure that the device will only consume energy from the structure, as follows.

According Newton's 3rd law of motion, the damping force acts simultaneously on the damper and the structure in opposite directions. Though, $\dot{z}_d$ is identical due to geometric compatibility conditions. Assume that at the damper we have $w(t)\dot{z}_d(t) \geq 0$ for all t. This power will be referred to as the *damper power*. It follows that for the structure we have $w(t)\dot{z}_d(t) \leq 0$ for all t. In other words, the mechanical power of the damping force, which acts on the structure, is always negative. This power will be referred to as the *control power*.

The fact that the damper power is positive is clearly reflected by each of the above mentioned damping force models. In the case of VD the damper power is:

$$w(\dot{z}_d(t))\dot{z}_d(t) = c_d|\dot{z}_d(t)|^\alpha \operatorname{sign}(\dot{z}_d(t))\dot{z}_d(t) = c_d|\dot{z}_d(t)|^\alpha |\dot{z}_d(t)| \geq 0$$

For MRD:

$$\begin{aligned} w(\dot{z}_d(t), i_c(t))\dot{z}_d(t) &= (c_d\dot{z}_d(t) + w^y(i_c(t))\operatorname{sign}(\dot{z}_d(t)))\dot{z}_d(t) \\ &= c_d\dot{z}_d(t)^2 + w^y(i_c(t))|\dot{z}_d(t)| \geq 0 \end{aligned}$$

For variable viscous damper:

$$w(\dot{z}_d(t), \theta(t))\dot{z}_d(t) = 2\sin(\theta(t))^2 c_d\dot{z}_d(t)^2 \geq 0$$

and for semi-active friction damper

$$w(\dot{z}_d(t), p(t))\dot{z}_d(t) = w^f(p(t))\operatorname{sign}(\dot{z}_d(t))\dot{z}_d(t) = w^f(p(t))|\dot{z}_d(t)| \geq 0$$

A non-positive control power assures that the control force will not excite the structure. Additionally, if it is negative then it will reduce the mechanical energy in the plant. That fact is illustrated by the next, elemental lemma and its remarks. It follows [63] and provides a sufficient condition for a dissipative control force in a multi-DOF case.

Lemma 2.1
Consider the mechanical plant given in Eq. (2.4) and let $E : \mathbb{R}^{n_z} \times \mathbb{R}^{n_z} \to [0, \infty)$ be

$$E(\mathbf{z}(t), \dot{\mathbf{z}}(t)) \triangleq \frac{1}{2}\mathbf{z}(t)^T \mathbf{K}\mathbf{z}(t) + \frac{1}{2}\dot{\mathbf{z}}(t)^T \mathbf{M}\dot{\mathbf{z}}(t)$$

i.e., the mechanical energy in the plant for some displacements and velocities in the DOFs. Let $\boldsymbol{\psi}_i$ be the i-th column of $\boldsymbol{\Psi}$. If $w_i(t)\boldsymbol{\psi}_i^T\dot{\mathbf{z}}(t) \leq 0$ for all $i \in [1, n_w]$ then $\dot{E}(\mathbf{z}(t), \dot{\mathbf{z}}(t)) \leq 0$.

Proof 2.1   Assume that $w_i(t)\boldsymbol{\psi}_i^T\dot{\mathbf{z}}(t) \leq 0$ for all $i \in [1, n_w]$. By substituting Eq. (2.4) into the full derivative of E, it is given that

$$\frac{\mathrm{d}}{\mathrm{d}t}E(\mathbf{z}(t), \dot{\mathbf{z}}(t)) = \mathbf{z}(t)^T\mathbf{K}\dot{\mathbf{z}}(t) + \dot{\mathbf{z}}(t)^T\mathbf{M}\ddot{\mathbf{z}}(t)$$

$$= \mathbf{z}(t)^T\mathbf{K}\dot{\mathbf{z}}(t) + \dot{\mathbf{z}}(t)^T(-\mathbf{K}\mathbf{z}(t) - \mathbf{C}_d\dot{\mathbf{z}}(t) + \boldsymbol{\Psi}\mathbf{w}(t))$$

$$= -\dot{\mathbf{z}}(t)^T\mathbf{C}_d\dot{\mathbf{z}}(t) + \dot{\mathbf{z}}(t)^T\boldsymbol{\Psi}\mathbf{w}(t))$$

As $\mathbf{C}_d \geq 0$ and $\dot{\mathbf{z}}(t)^T\boldsymbol{\Psi}\mathbf{w}(t) = \sum_{i=1}^{n_w} w_i(t)\boldsymbol{\psi}_i^T\mathbf{z}(t) \leq 0$ then $\dot{E}(\mathbf{z}(t), \dot{\mathbf{z}}(t)) \leq 0$.

Remarks.

■ It can be shown that $\boldsymbol{\psi}_i^T\dot{\mathbf{z}}(t) = \dot{z}_d(t)$ (See [31], pp. 95–96). Hence $w_i(t)\boldsymbol{\psi}_i^T\dot{\mathbf{z}}(t) \leq 0$ implies non-positive control power.

■ If $w_i(t)\boldsymbol{\psi}_i^T\dot{\mathbf{z}}(t) \leq 0$ for all $t \in [0, t_f]$ then the work done by the control force is always non-positive, i.e., the control force cannot excite the plant.

Note that even though one can find analogies between them, the dissipativeness discussed here is different from the general concept of dissipative systems, which is known from general control theory and is discussed in the next section.

2.3   Dissipative Systems

As it was mentioned above, an important property of linear structural models, which are controlled by viscous fluid and/or semi-active dampers, is that their stability is guaranteed by the physical dissipativeness of these devices. This property is related to the general concept of dissipative systems, as follows.

At the 1960s, a class of dynamic models, characterized by some input-output property, began to attract the attention of many researchers. The theory of these models has taken inspiration from physical systems that contain no internal energy source. The models were said to be *dissipative*, and when some additional conditions were satisfied they were said to be *passive*. The properties of these models have significant implications regarding the stability of the controlled plant [101]. They also provide a basis for several control design methods [10, 32], especially for adaptive systems [13, 50].

In structural control, the theory of dissipative systems is of great importance as its implications on stability are related to the structure's residents safety. However, a problem that arises when the structural control designer is willing to use results from this field, is that although they use the same terms, the definitions of passive and dissipative systems in structural control are rather different than those used in general control theory. In this section, several parts from dissipative and passive systems theory, in the general control sense, are given and then related to dissipativity and passivity, in the structural control sense. Similarities and dissimilarities are highlighted.

Consider the state space model from Eqs. (2.1) and (2.2). Let a *supply rate* be a function $\sigma : \mathbb{R}^{n_u} \times \mathbb{R}^{n_y} \to \mathbb{R}$ such that $\int_{t_0}^{t_1} |\sigma(\mathbf{u}(t), \mathbf{y}(t))| \mathrm{d}t$ is finite for any interval $[t_0, t_1] \in [0, t_f]$, and for any input and output trajectories $\mathbf{u}$ and $\mathbf{y}$, respectively [101].

Definition 2.1 (Dissipative Model) [101]   Let a state space model be defined by Eqs. (2.1) and (2.2), and let $\mathbf{u}$, $\mathbf{x}$, $\mathbf{y}$ be its input, state and output trajectories, respectively. This model is said to be *dissipative* with respect to σ, if there exists a nonnegative function $S : \mathbb{R}^n \to [0, \infty)$ such that

$$S(\mathbf{x}(t_0)) + \int_{t_0}^{t_1} \sigma(\mathbf{u}(t), \mathbf{y}(t)) \mathrm{d}t \geq S(\mathbf{x}(t_1))$$

for any $0 \leq t_0 \leq t_1 \leq t_f$, initial state $\mathbf{x}(0) \in \mathbb{R}^n$ and input $\mathbf{u} \in \{\mathbb{R} \to \mathbb{R}^{n_u}\}$.

S is denoted as *storage function*. From the physical viewpoint, σ refers to the input power and S refers to the plant's energy. Consequently, the meaning of Definition 2.1 is that a system is dissipative if the increase in its energy (S) during the interval $[t_0, t_1] \subseteq [0, t_f]$ is no greater than the energy supplied, i.e., $\int_{t_0}^{t_1} \sigma(\mathbf{u}(t), \mathbf{y}(t)) \mathrm{d}t \geq S(\mathbf{x}(t_1)) - S(\mathbf{x}(t_0))$.

If S is differentiable then one can require $\dot{S}(\mathbf{x}(t)) \leq \sigma(\mathbf{u}(t), \mathbf{y}(t))$ for all $t \in (t_0, t_1)$.

The notation $\mathbf{y}(t, \mathbf{u}, \xi, t_0)$ will be used to denote an output trajectory that corresponds to an input $\mathbf{u}$ and an initial condition $\mathbf{x}(t_0) = \xi$.

Definition 2.2 (Available storage) [101]   Let $\xi \in \mathbb{R}^n$. The *available storage*, $S_a : \mathbb{R}^n \to [0,\infty)$ is defined by

$$S_a(\xi) \triangleq \sup_{\substack{\tau \in [t_0, t_f] \\ \mathbf{u} \in \{\mathbb{R} \to \mathbb{R}^{n_u}\}}} \left(-\int_{t_0}^{\tau} \sigma(\mathbf{u}(t), \mathbf{y}(t, \mathbf{u}, \xi, t_0)) dt \right)$$

The fact that $S_a(\xi) \geq 0$ is clear as the supremum's domain contains 0. It is common to interpret $S_a(\xi)$ as the maximum energy that can be extracted from the system due to initial condition ξ. An interesting property of S_a is given in the following lemma.

Lemma 2.2 [101]
Let $\tilde{\mathbf{u}}$, $\tilde{\mathbf{x}}$ and $\tilde{\mathbf{y}}$ be trajectories that satisfy Eqs. (2.1) and (2.2). Then

$$S_a(\tilde{\mathbf{x}}(t_0)) + \int_{t_0}^{t_1} \sigma(\tilde{\mathbf{u}}(t), \tilde{\mathbf{y}}(t)) dt \geq S_a(\tilde{\mathbf{x}}(t_1))$$

for any $0 \leq t_0 \leq t_1 \leq t_f$.

The following proof is essentially a rigorous formulation of the proof outline given in [101].

Proof 2.2   Let $V \subset \{[0, t_f] \to \mathbb{R}^{n_u}\}$ be a set of trajectories that satisfy

$$\mathbf{v}(t) \begin{cases} = \tilde{\mathbf{u}}(t) & t \in [t_0, t_1] \\ \in \mathbb{R}^{n_u} & \text{otherwise} \end{cases}$$

for any $\mathbf{v} \in V$. In other words, all the trajectories in V coincide with $\tilde{\mathbf{u}}$ at the interval $[t_0, t_1]$. Outside this interval the trajectories can be any function from $\{\mathbb{R} \to \mathbb{R}^{n_u}\}$. It follows that

$$\sup_{\substack{\tau \in [t_1, t_f] \\ \mathbf{u} \in V}} \left(-\int_{t_0}^{\tau} \sigma(\mathbf{u}(t), \mathbf{y}(t, \mathbf{u}, \tilde{\mathbf{x}}(t_0), t_0)) dt \right)$$

$$= \sup_{\substack{\tau \in [t_1, t_f] \\ \mathbf{u} \in V}} \left(-\int_{t_0}^{t_1} \sigma(\tilde{\mathbf{u}}(t), \mathbf{y}(t, \tilde{\mathbf{u}}, \tilde{\mathbf{x}}(t_0), t_0)) dt - \int_{t_1}^{\tau} \sigma(\mathbf{u}(t), \mathbf{y}(t, \mathbf{u}, \tilde{\mathbf{x}}(t_0), t_0)) dt \right)$$

$$= -\int_{t_0}^{t_1} \sigma(\tilde{\mathbf{u}}(t), \mathbf{y}(t, \tilde{\mathbf{u}}, \tilde{\mathbf{x}}(t_0), t_0)) dt + \sup_{\substack{\tau \in [t_1, t_f] \\ \mathbf{u} \in \{\mathbb{R} \to \mathbb{R}^{n_u}\}}} \left(-\int_{t_1}^{\tau} \sigma(\mathbf{u}(t), \mathbf{y}(t, \mathbf{u}, \tilde{\mathbf{x}}(t_0), t_0)) dt \right)$$

because the first integral is identical for all the trajectories in V. As $\tilde{\mathbf{x}}$ is the dynamic state, for any $t \geq t_1$ and $\mathbf{u} \in V$ we have $\mathbf{y}(t,\mathbf{u},\tilde{\mathbf{x}}(t_0),t_0) = \mathbf{y}(t,\mathbf{u},\tilde{\mathbf{x}}(t_1),t_1)$. Therefore

$$
\sup_{\substack{\tau \in [t_1,t_f] \\ \mathbf{u} \in V}} \left(-\int_{t_0}^{\tau} \sigma(\mathbf{u}(t),\mathbf{y}(t,\mathbf{u},\tilde{\mathbf{x}}(t_0),t_0))\mathrm{d}t \right) = -\int_{t_0}^{t_1} \sigma(\tilde{\mathbf{u}}(t),\mathbf{y}(t,\tilde{\mathbf{u}},\tilde{\mathbf{x}}(t_0),t_0))\mathrm{d}t
$$

$$
+ \sup_{\substack{\tau \in [t_1,t_f] \\ \mathbf{u} \in \{\mathbb{R}\to\mathbb{R}^{n_u}\}}} \left(-\int_{t_1}^{\tau} \sigma(\mathbf{u}(t),\mathbf{y}(t,\mathbf{u},\tilde{\mathbf{x}}(t_1),t_1))\mathrm{d}t \right)
$$

$$
= -\int_{t_0}^{t_1} \sigma(\tilde{\mathbf{u}}(t),\mathbf{y}(t,\mathbf{u},\tilde{\mathbf{x}}(t_0),t_0))\mathrm{d}t + S_a(\tilde{\mathbf{x}}(t_1))
$$

Recall that if $A \subset B$ then $\sup A \leq \sup B$. By virtue of S_a's definition:

$$
S_a(\tilde{\mathbf{x}}(t_0)) = \sup_{\substack{\tau \in [t_0,t_f] \\ \mathbf{u} \in \{\mathbb{R}\to\mathbb{R}^{n_u}\}}} \left(-\int_{t_0}^{\tau} \sigma(\mathbf{u}(t),\mathbf{y}(t,\mathbf{u},\tilde{\mathbf{x}}(t_0),t_0))\mathrm{d}t \right)
$$

$$
\geq \sup_{\substack{\tau \in [t_0,t_f] \\ \mathbf{u} \in V}} \left(-\int_{t_0}^{\tau} \sigma(\mathbf{u}(t),\mathbf{y}(t,\mathbf{u},\tilde{\mathbf{x}}(t_0),t_0))\mathrm{d}t \right)
$$

$$
\geq \sup_{\substack{\tau \in [t_1,t_f] \\ \mathbf{u} \in V}} \left(-\int_{t_0}^{\tau} \sigma(\mathbf{u}(t),\mathbf{y}(t,\mathbf{u},\tilde{\mathbf{x}}(t_0),t_0))\mathrm{d}t \right)
$$

$$
= -\int_{t_0}^{t_1} \sigma(\tilde{\mathbf{u}}(t),\mathbf{y}(t,\tilde{\mathbf{u}},\tilde{\mathbf{x}}(t_0),t_0))\mathrm{d}t + S_a(\tilde{\mathbf{x}}(t_1))
$$

and the proof is completed.

It was proved [101] that a system is dissipative if and only if the available storage S_a is finite. Moreover, if it is finite then S_a is a storage function with respect to σ and it is a lower bound for all storage functions S, i.e., $S(\xi) \geq S_a(\xi) \geq 0$ [101]. That allows to determine dissipativity of a system by studying S_a rather than investigating the existence of a storage function S.

In general control theory, passive systems are a subclass of dissipative systems, defined as follows.

Definition 2.3 (Passive Model) [20]   Let a state space model be defined by Eqs. (2.1) and (2.2), and let **u**, **x**, **y** be its input, state and output trajectories, respectively. This model is said to be *passive* if (1) $n_y = n_u$, (2) it is dissipative with respect to $\sigma(\mathbf{u}(t), \mathbf{y}(t)) = \mathbf{u}(t)^T \mathbf{y}(t)$ and (3) the corresponding storage function satisfies $S(\mathbf{0}) = 0$.

A famous theorem, which is related to passive models, is the Kalman-Yakubobich-Popov (KYP) Lemma [102]. It is equivalent to Definition 2.3 only that it refers to the state space representation rather than the input-output one and it uses an algebraic condition instead of an integral formulation. It is given below with relation to an LTI systems with $\mathbf{D} = \mathbf{0}$.

Lemma 2.3 (KYP Condition) [102]
Consider an LTI model that satisfies Eqs. (2.5) and (2.6) with $n_y = n_u$ and $\mathbf{D} = \mathbf{0}$. It is passive if and only if there exists a positive definite $\mathbf{P} \in \mathbb{R}^{n \times n}$ that satisfies $\mathbf{P}\mathbf{A} + \mathbf{A}^T \mathbf{P} = -\mathbf{Q}$ and $\mathbf{P}\mathbf{B} = \mathbf{C}^T$ for some $\mathbf{Q} \geq 0$.

Remarks.

■ Formulation of this lemma for linear time varying (LTV) systems can be found in [32].

■ The concept of dissipativity defined here is more general than the physical dissipativity related to semi-active and passive dampers (Section 2.2). The former is related to σ and S that may have no physical meaning, while the latter has obvious physical meaning.

■ While in structural control semi-active and passive dampers are considered different, in the general control theory both of them fall in the category of passive systems, as follows. Let a dynamic model of physical dissipative control device be:

$$\dot{\mathbf{x}}_d(t) = \mathbf{f}_d(t, \mathbf{x}_d(t), u(t))$$
$$y(t) = \mathbf{h}_d(t, \mathbf{x}_d(t), u(t))$$

where the input $u = \dot{z}_d$ is the device's elongation rate and the output $y = w$ is the device's tensile force. The physical dissipativity of the device guarantees that $u(t)y(t) = \dot{z}_d(t)w(t) \geq 0$. Hence by choosing $S(\xi) \triangleq 0$, the condition in Definition 2.1 holds and as $S(0) = 0$ this system is passive. Hence, semi-active dampers are passive systems, in the general control theory sense, with respect to $u = \dot{z}_d$ and $y = w$.

Note that input current, which is used to alter the semi-active damper properties, is not considered here as an input, but rather as external disturbance signal. It makes sense since the input-output representation, which is addressed by the passivity theory, considers **u** and **y** that are used to interconnect the system with other systems (see Theorem 2.1). In semi-active controllers, the physical setup defines the closed loop with respect to the damper's velocity input while the input current does not play a role in this discussion.

The block diagram in Fig. 2.4(a) describes a closed loop that consists of two interconnected dynamic systems - $H_1 : \mathbf{u}_1 \to \mathbf{y}_1$ and $H_2 : \mathbf{u}_2 \to \mathbf{y}_2$. Closed loop stability is a major concern in control theory. It is well known that the stability of H_1 and H_2 is not enough to ensure the stability of the entire closed loop[1]. However, the situation is different when dealing passive systems.

Consider a system defined by Eqs. (2.1) and (2.2). Let $\mathbf{x}(t, \mathbf{u}, \boldsymbol{\xi}, t_0)$ and $\mathbf{y}(t, \mathbf{u}, \boldsymbol{\xi}, t_0)$ be its state and output trajectories, respectively, due to control trajectory $\mathbf{u}$ and initial condition $\mathbf{x}(t_0) = \boldsymbol{\xi}$. $t_0 \leq t$. The system is *zero state detectable* (ZSD) if for any $\boldsymbol{\xi} \in \mathbb{R}^n$, $\mathbf{y}(t, 0, \boldsymbol{\xi}, t_0) = 0$ for all $t_0 \leq t$ implies that $\lim_{t \to \infty} \mathbf{x}(t, 0, \boldsymbol{\xi}, t_0) = 0$ [10].

Theorem 2.1 [10]
Let $H_1 : \mathbf{u}_1 \to \mathbf{y}_1$ and $H_2 : \mathbf{u}_2 \to \mathbf{y}_2$ be two passive systems. If H_1 and H_2 are interconnected in a feedback form, as in Fig. 2.4(a), then the closed loop is passive with respect to the input **r** *and the output* **y**. *Additionally, if H_1 and H_2 are ZSD, then the origin $(\mathbf{x}_1(0), \mathbf{x}_2(0)) = (\mathbf{0}, \mathbf{0})$ is stable.*

If the damper's dynamics are neglected, as in common viscous damping models (Section 2.2), then the damper's output force is given by some sector bounded static mapping, rather than a passive dynamic system. The closed loop for this case is given by Fig. 2.4(b). This common situation was discussed in [32] for LTV systems and in Section 4 of [109] for a more general case. Under suitable conditions, closing the loop with such sector bounded mapping ensures closed loop stability, as in interconnected passive systems [109, 32].

The next lemma shows that a state space model of a controlled structure is passive with respect to the control forces and the relative velocities of the actuators anchors.

---

[1]One can take for example transfer functions of two systems—$\frac{-1}{s+1}$ and $\frac{1}{s+1}$. The transfer function of the closed loop is $\frac{-s-1}{s(s+2)}$ which is unstable.

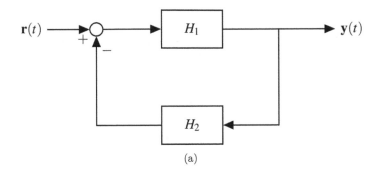

(a)

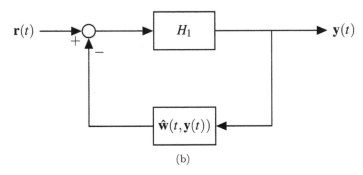

(b)

Figure 2.4: Interconnected systems.

Lemma 2.4
The LTI system, defined in Eqs. (2.5) and (2.6) with $\mathbf{C} = \begin{bmatrix} \mathbf{0} & \mathbf{\Psi}^T \end{bmatrix}$ and $\mathbf{D} = \mathbf{0}$, is passive.

Proof 2.3   Let

$$\mathbf{P} = \begin{bmatrix} \mathbf{K} & \mathbf{0} \\ \mathbf{0} & \mathbf{M} \end{bmatrix} \tag{2.12}$$

then

$$\mathbf{PA} + \mathbf{A}^T\mathbf{P} = \begin{bmatrix} \mathbf{K} & \mathbf{0} \\ \mathbf{0} & \mathbf{M} \end{bmatrix} \begin{bmatrix} \mathbf{0} & \mathbf{I} \\ -\mathbf{M}^{-1}\mathbf{K} & -\mathbf{M}^{-1}\mathbf{C}_d \end{bmatrix} + \begin{bmatrix} \mathbf{0} & -\mathbf{K}\mathbf{M}^{-1} \\ \mathbf{I} & -\mathbf{C}_d\mathbf{M}^{-1} \end{bmatrix} \begin{bmatrix} \mathbf{K} & \mathbf{0} \\ \mathbf{0} & \mathbf{M} \end{bmatrix}$$

$$= -\begin{bmatrix} \mathbf{0} & \mathbf{0} \\ \mathbf{0} & 2\mathbf{C}_d \end{bmatrix} \leq 0$$

$$\mathbf{PB} = \begin{bmatrix} \mathbf{K} & \mathbf{0} \\ \mathbf{0} & \mathbf{M} \end{bmatrix} \begin{bmatrix} \mathbf{0} \\ \mathbf{M}^{-1}\mathbf{\Psi} \end{bmatrix} = \begin{bmatrix} \mathbf{0} \\ \mathbf{\Psi} \end{bmatrix} = \mathbf{C}^T$$

Hence the system is passive by virtue of KYP lemma.

Remarks.

- Let $S(\mathbf{x}(t)) \triangleq \frac{1}{2}\mathbf{x}(t)^T\mathbf{P}\mathbf{x}(t)$ where $\mathbf{x}$ is defined by Eq. (2.5) and $\mathbf{P}$ is defined by Eq. (2.12). It is clear that S is identical to E from Lemma 2.1.

- According the first remark to Lemma 2.1, if $\boldsymbol{\Psi}$ is the control force distribution matrix for a set of control devices, then $\dot{z}_d(t) = \begin{bmatrix} \mathbf{0} & \boldsymbol{\Psi}^T \end{bmatrix}\mathbf{x}(t)$ is the elongation rate of these devices.

- Figure 2.5 provides an example for the physical setup of a damper in a structure and its relation with the close loop from Fig. 2.4(a). Figures 2.5(a) and 2.5(b) describe the input-output relation of the plant and its physical interpretation. The left drawing in Fig. 2.5(c) illustrates a simple damper's setup. The damper is connected to the roof beam by a stiff Chevron brace [88] such that the plant motion is coupled with that of the damper. Hence $z_d \equiv z$. Though, the damping force is transferred to the plant by the braces with an opposite sign due to Newton's 3rd law of motion. That can be verified by neglecting the brace mass and applying Newton's 2nd law in horizontal direction, as illustrated in the right drawing in Fig. 2.5(c). Hence, $w_1 \equiv -w$, which means negative feedback. As Lemma 2.4 assures that the structure is passive in $(w_1, \dot{z})$ and as the damper physical dissipativity assures that it is passive in $(w, \dot{z}_d)$, then the closed loop is passive in $(r, \dot{z})$. If the plant and the damper are ZSD, then the closed loop is stable. As these properties (passivity and ZSD) are inherent, it follows that the closed loop stability is inherent too.

Stability of LTI plants is often analyzed by the plant's stability margin, i.e., the distance of its eigenvalues from the imaginary axis. Consider a set of VDs, which are modeled by Eq. (2.10). The inherent stability of a plant, controlled by these VDs, is reflected by an improvement in the plant's stability margin, as follows.

The free response of a plant, controlled by this set of VDs, is governed by

$$\dot{\mathbf{x}}(t) = \left(\mathbf{A} - \sum_{i=1}^{n_u} \mathbf{b}_i \tilde{u}_i \mathbf{c}_i \right) \mathbf{x}(t); \quad \mathbf{x}(0), t \in (0, \infty)$$

Define $\mathbf{A}_{cl} : \mathbb{R}^{n_u} \rightarrow \mathbb{R}^{n \times n}$ as:

$$\mathbf{A}_{cl}(\tilde{\mathbf{u}}) \triangleq \mathbf{A} - \sum_{i=1}^{n_u} \mathbf{b}_i \tilde{u}_i \mathbf{c}_i$$

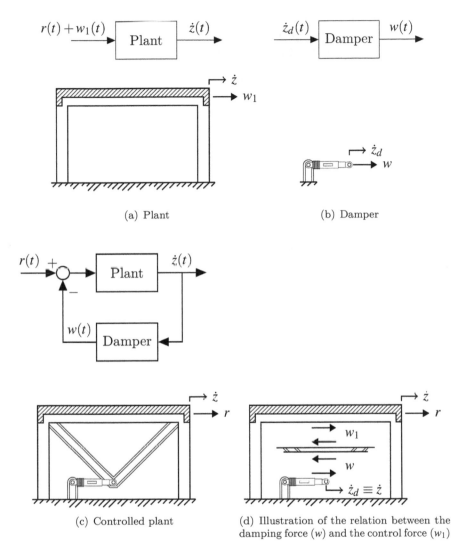

Figure 2.5: A controlled frame of a structure.

The derivative of some eigenvalue of $\mathbf{A}_{cl}$ at $\tilde{u}_i$ is [81]:

$$\frac{\partial}{\partial \tilde{u}_i} \lambda_i = \frac{-\mathbf{v}_{li} \mathbf{b}_i \mathbf{c}_i \mathbf{v}_{li}}{\mathbf{v}_{li} \mathbf{v}_{ri}}$$

where $\mathbf{v}_{ri}$ and $\mathbf{v}_{li}$ are the right and left eigenvectors of $\mathbf{A}$, respectively. Assume, without loss of generality, that the eigenvectors are normalized such that $\mathbf{v}_{li}\mathbf{v}_{ri} = 1$. It implies that as long that

$$\mathrm{Re}(\mathbf{v}_{li}\mathbf{b}_i\mathbf{c}_i\mathbf{v}_{ri}) > 0$$

the stability margin of eigenvalue λ_i will increase, i.e., the feedback will move that eigenvalue towards the left side of the complex plain.

Chapter 3

Optimal Control

Usually, when trying to solve practical control problems, there are more than one solution available. The variety of suitable solutions complicates the design process as it increases the amount of dilemmas, facing the control designer and, as a result, the question "what is the optimal solution?" naturally arises. This issue becomes more complicated when the systems involved are large and complex. The sub-field of control theory that addresses this problem is known as *optimal control theory*. This chapter sums-up several known theoretical results in optimal control theory.

Before the search after optimal control design begins, a clear answer to the question "How a control solution's optimality can be evaluated?" should be provided. To answer that, the optimal control design starts with a definition of some quantitative criterion that provides an evaluation for each of the candidate solutions. This criterion, denoted here as *performance index*, allows evaluating each design and comparing it with others. In optimal structural control, the performance index is usually a real, non-negative, quantitative measure of the performance, where superior performance yields a smaller numerical value [18]. It follows that the optimal control solution is the one with minimal cost.

Let

$$\dot{\mathbf{x}}(t) = \mathbf{f}(t, \mathbf{x}(t), \mathbf{u}(t)); \quad \mathbf{x}(0), \forall t \in (0, t_f) \tag{3.1}$$

be the state equation; $\mathscr{U} \subseteq \{\mathbb{R} \to \mathbb{R}^{n_u}\}$ be a set of admissible control trajectories and $\mathscr{X} \subseteq \{\mathbb{R} \to \mathbb{R}^n\}$ be a set of state trajectories that are reachable from $\mathscr{U}$ and $\mathbf{x}(0)$. In what follows, the term *process* will refer to some pair

($\mathbf{x} \in \mathscr{X}$, $\mathbf{u} \in \mathscr{U}$). A pair $(\mathbf{x}, \mathbf{u})$ that satisfies Eq. (3.1) is called an *admissible process*. Let the performance index be a non-negative functional

$$J : \mathscr{X} \times \mathscr{U} \to [0, \infty)$$

$\mathbf{f}$ and J define an optimal control problem, as follows.

Definition 3.1 The pair $(\mathbf{x}^*, \mathbf{u}^*)$ is said to be an optimal admissible process if it satisfies the following equations:

$$\dot{\mathbf{x}}^*(t) = \mathbf{f}(t, \mathbf{x}^*(t), \mathbf{u}^*(t)) \quad \mathbf{x}(0), \forall t \in (0, t_f) \tag{3.2}$$

$$(\mathbf{x}^*, \mathbf{u}^*) = \arg \min_{\substack{\mathbf{x} \in \mathscr{X} \\ \mathbf{u} \in \mathscr{U}}} J(\mathbf{x}, \mathbf{u}) \tag{3.3}$$

A general form for J, which is commonly used for evaluating the performance of some controlled plant over the time interval $t \in [0, t_f]$, is the functional

$$J(\mathbf{x}, \mathbf{u}) = l_f(\mathbf{x}(t_f)) + \int_0^{t_f} l(t, \mathbf{x}(t), \mathbf{u}(t)) \mathrm{d}t \tag{3.4}$$

where $l_f : \mathbb{R}^n \to [0, \infty)$ is the *terminal cost*. That is, the contribution of $\mathbf{x}(t_f)$ to the evaluated performance; $l : \mathbb{R} \times \mathbb{R}^n \times \mathbb{R}^{n_u} \to [0, \infty)$ is the *process cost* and its integral evaluates the performance of $(\mathbf{x}, \mathbf{u})$ during the control time.

In structural control most of the problems are concerned only with reducing the process cost while the terminal cost has no significance. This type of problem uses the same performance index as before but without the terminal cost:

$$J(\mathbf{x}, \mathbf{u}) = \int_0^{t_f} l(t, \mathbf{x}(t), \mathbf{u}(t)) \mathrm{d}t \tag{3.5}$$

Additionally, for many structural control problems, the control duration is not an issue. In that case it is popular to use infinite horizon by taking $t_f \to \infty$.

Another type of optimal control problems is *parametric optimal control*. In these problems, the controller configuration is determined apriori and the goal is to minimize some performance index by changing the control parameters, rather than control trajectories. Following the above terminology, the goal is to find $(\mathbf{x}^*, \mathbf{u}^*)$ where $\mathbf{u} \in \mathscr{U}$ and $\mathscr{U}$ is the set of constant control trajectories, that is $\mathscr{U} = \{u | u(t) = \tilde{u} \in \mathbb{R}\}$. A corresponding performance index will be:

$$J(\mathbf{x}, \mathbf{u}) = J(\mathbf{x}, \tilde{\mathbf{u}}) = \int_0^{t_f} l(\mathbf{x}(t), \tilde{\mathbf{u}}) dt + l_{st}(\tilde{\mathbf{u}})$$

where $\tilde{\mathbf{u}} \in \mathbb{R}^{n_u}$ is the parameters vector and $l_{st} : \mathbb{R} \times \mathbb{R}^{n_u} \to [0, \infty)$ is some 'static' parameters weighting.

3.1 Lagrange's Multipliers

A fundamental optimal control problem, is to find the optimal control trajectories when no constraints are imposed on the desired optimum. A common approach to solve this problem is to formulate the state equation as an equality constraint and then to use Lagrange's multipliers method.

This well known method starts with definition of an augmented performance index:

$$J^a(\mathbf{x}, \mathbf{u}, \mathbf{p}) = l_f(\mathbf{x}(t_f)) + \int_0^{t_f} \left(H(t, \mathbf{x}(t), \mathbf{u}(t), \mathbf{p}(t)) - \mathbf{p}(t)^T \dot{\mathbf{x}}(t) \right) dt \qquad (3.6)$$

where

$$H(t, \mathbf{x}(t), \mathbf{u}(t), \mathbf{p}(t)) \triangleq l(t, \mathbf{x}(t), \mathbf{u}(t)) + \mathbf{p}(t)^T \mathbf{f}(t, \mathbf{x}(t), \mathbf{u}(t)) \qquad (3.7)$$

is known as the *Hamiltonian* and $\mathbf{p} : \mathbb{R} \to \mathbb{R}^n$ is a vector function of Lagrange's multipliers, also known as *costate*. According to the calculus of variations, the gradient of J^a should vanish at an admissible optimum [53]. This necessary condition leads to the following theorem.

Theorem 3.1 [53]
Let an optimal control problem be defined by the following state equation, performance index and set of control trajectories:

$$\dot{\mathbf{x}}(t) = \mathbf{f}(t, \mathbf{x}(t), \mathbf{u}(t)); \quad \mathbf{x}(0), \forall t \in (0, t_f) \qquad (3.8a)$$

$$J(\mathbf{x}, \mathbf{u}) = l_f(\mathbf{x}(t_f)) + \int_0^{t_f} l(t, \mathbf{x}(t), \mathbf{u}(t)) dt \qquad (3.8b)$$

$$\mathbf{u} \in \{\mathbb{R} \to \mathbb{R}^{n_u}\} \qquad (3.8c)$$

If $(\mathbf{x}^*, \mathbf{u}^*)$ *is an admissible optimum then it satisfies the following equations*

$$\dot{\mathbf{x}}^*(t) = \mathbf{f}(t, \mathbf{x}^*(t), \mathbf{u}^*(t)); \quad \mathbf{x}^*(0) = \mathbf{x}(0) \tag{3.9a}$$

$$\dot{\mathbf{p}}(t) = -\mathbf{f_x}(t, \mathbf{x}^*(t), \mathbf{u}^*(t))^T \mathbf{p}(t) - l_{\mathbf{x}}(t, \mathbf{x}^*(t), \mathbf{u}^*(t))^T;$$
$$\mathbf{p}(t_f) = l_{f\mathbf{x}}(\mathbf{x}_k(t_f))^T \tag{3.9b}$$

$$0 = -\mathbf{p}(t)^T \mathbf{f_u}(t, \mathbf{x}^*(t), \mathbf{u}^*(t)) - l_{\mathbf{u}}(t, \mathbf{x}^*(t), \mathbf{u}^*(t)) \tag{3.9c}$$

The solution of these equations allows finding a *candidate* admissible optimum. Here the term *candidate* is important because satisfying these equations is merely a necessary condition rather than a sufficient one.

This approach forms a theoretical basis for many optimal control solutions. For example, an optimal controller law for bilinear system and a quadratic cost for tracking problem, was suggested [62]. The theoretical framework was constructed by using Lagrange's multipliers and an iterative algorithm was suggested for its solution; A numerical algorithm for solving linear quadratic regulator problem with constraints on the state and the input was developed by using Lagrange's multipliers [47].

3.2   Pontryagin's Minimum Principle

Sometimes physical considerations set limitations on the control trajectories. Such problems can be found in many real life control problems, such as control of mechanical plants [94, 17, 41], quantum mechanics [57] and industrial processes [47]. From optimization viewpoint, an optimal design that ignores these constraints, might derive $\mathbf{u}$ that is impractical and therefore irrelevant. Hence, these control constraints should be taken into account during the optimization. However, incorporating constraints in a control problem usually poses a significant challenge to the control designer. For example, it is known that imposing constraints on *linear quadratic* (LQ) problems, which are common in optimal control design, fundamentally impacts their solvability [30].

An important property of $\mathscr{U}$, which is defined by the given constraints, is its being closed or open relative to $\mathbb{R}^{n_u}$. When $\mathscr{U}$ is closed, variational methods, such as Lagrange's multipliers, are not appropriate because the gradient of H with respect to $\mathbf{u}$ is undefined on $\mathscr{U}$'s boundaries. A famous theorem in control optimization, which can be used in such case, is Pontryagin's minimum principle. It provides necessary conditions for an admissible process, where $\mathbf{u}^* \in \mathscr{U}$ and $\mathscr{U}$ can be either open or closed.

Theorem 3.2 [53]
*Let a constrained control optimization problem be defined by the following state
equation, performance index and set of admissible control trajectories:*

$$\dot{\mathbf{x}}(t) = \mathbf{f}(t, \mathbf{x}(t), \mathbf{u}(t)); \quad \mathbf{x}(0), \forall t \in (0, t_f) \tag{3.10a}$$

$$J(\mathbf{x}, \mathbf{u}) = l_f(\mathbf{x}(t_f)) + \int_0^{t_f} l(t, \mathbf{x}(t), \mathbf{u}(t)) \mathrm{d}t \tag{3.10b}$$

$$\mathbf{u} \in \mathscr{U} \tag{3.10c}$$

If $(\mathbf{x}^*, \mathbf{u}^*)$ is an admissible optimum then it satisfies the following equations

$$\dot{\mathbf{x}}^*(t) = \mathbf{f}(t, \mathbf{x}^*(t), \mathbf{u}^*(t)); \quad \mathbf{x}^*(0) = \mathbf{x}(0) \tag{3.11a}$$

$$\dot{\mathbf{p}}(t) = -\mathbf{f_x}(t, \mathbf{x}^*(t), \mathbf{u}^*(t))^T \mathbf{p}(t) - l_\mathbf{x}(t, \mathbf{x}^*(t), \mathbf{u}^*(t))^T; \\ \mathbf{p}(t_f) = l_{f\mathbf{x}}(\mathbf{x}_k(t_f))^T \tag{3.11b}$$

$$H(t, \mathbf{x}^*(t), \mathbf{u}^*(t), \mathbf{p}(t)) \le H(t, \mathbf{x}^*(t), \mathbf{u}(t), \mathbf{p}(t)) \quad \forall \mathbf{u} \in \mathscr{U}, t \in [0, t_f] \tag{3.11c}$$

where H is the Hamiltonian function that was defined in Eq. (3.7).

Similarly to Lagrange's multipliers method, this theorem provides a necessary condition, hence the calculated trajectory is merely a candidate optimum.

Pontryagin's minimum principle was used in many optimal control problems. For example, for computing optimal open-loop trajectories of systems with state and control inequality constraints, a quadratic performance index and linear state's dynamics [46].

One of the first contributions to optimal control of bilinear systems [14] was formulated by Pontryagin's minimum principle [19]. The addressed problem was defined by single input, homogeneous-in-state, bilinear system; an infinite horizon quadratic performance index on the state and a bounded control signal. The problem was solved by Pontryagin's minimum principle and the control law took the form of a bang-bang control law.

A method for optimal control of bilinear systems with quadratic performance index was presented with emphasis on the finite horizon problem [48]. A suitable iterative solution method was formulated. A Bilinear-quadratic optimal control problem (BQR) was defined for an homogeneous bilinear system and quadratic performance index [1]. No constraints were imposed on the control. An iterative scheme that produces a linear control law was suggested. Its derivation was based on Pontryagin's minimum principle and the successive approximations approach. Essentially, the method requires solving the differential Lyapunov equation at each iteration and a proof of convergence was given.

Even though it was used in many control optimization problems, the efficiency of Pontryagin's minimum principle is more evident in optimal open-loop control problems [56].

3.3 Karush-Kuhn-Tucker Necessary Conditions

Some constrained optimal control problems consist of inequality constraints. The optimum in this case is constrained to a set of trajectories (possibly closed ones) whose boundaries are defined by the imposed inequalities. A theorem that defines necessary conditions for the optimal solution of this problem was found by a number of researchers—W. Karush, H. W. Kuhn, A. W. Tucker, F. John and others. It was published in 1950 by Kuhn-Tucker [58], however, they were not the first – it had been already proved in 1939 by W. Karush in his master thesis [54]. Today, it is common to refer this theorem as the 'Karush-Kuhn-Tucker theorem' (KKT). The main idea of the KKT theorem will be illustrated here by an inequality constrained, static optimization problem.

Let $\mathbf{v} \in \mathbb{R}^{n_v}$ be a vector of parameters. An inequality constrained optimization problem is defined by

$$J : \mathbb{R}^{n_v} \to [0, \infty) \tag{3.12a}$$

$$\{f_1^c, \ldots, f_m^c\} \tag{3.12b}$$

where each mapping $f_i^c : \mathbb{R}^{n_v} \to \mathbb{R}$ defines an inequality and J is a performance index. f_i^c and J are assumed to be smooth. An admissible optimum is a $\mathbf{v}^* = \arg\min_{\mathbf{v} \in \mathbb{R}^{n_v}} J(\mathbf{v})$ such that $f_i^c(\mathbf{v}^*) \leq 0 \forall i = 1, \ldots, m$.

Theorem 3.3 (KKT theorem) [15]
If $\mathbf{v}^*$ is an admissible optimum then there exists a vector $\boldsymbol{\mu}^* \in \mathbb{R}^m$ such that:

$$J_{\mathbf{v}}(\mathbf{v}^*) + (\boldsymbol{\mu}^*)^T \mathbf{f}_{\mathbf{v}}^c(\mathbf{v}^*) = \mathbf{0} \tag{3.13a}$$

$$\mu_i^* \begin{cases} \geq 0 & , f_i^c(\mathbf{v}^*) = 0 \\ = 0 & , f_i^c(\mathbf{v}^*) < 0 \end{cases} \tag{3.13b}$$

where $\mathbf{f}^c = (f_1^c, \ldots, f_m^c)$. Here $J_{\mathbf{v}}$ and $\mathbf{f}_{\mathbf{v}}^c$ are the gradient and jacobian matrix, respectively.

Hereinafter μ_i^* will be called *KKT multiplier*. From geometric viewpoint, there is an interesting relation between KKT's conditions and Lagrange's multipliers. Set aside, for a moment, the fact that $\{f_i^c\}$ are inequality constraints and assume that $f_i^c(\mathbf{v}^*) = 0$ for all $i = 1, \ldots, m$ and that $(\mu_i^*)_{i=1}^m$ is a vector of Lagrange multipliers. It follows that every f_i^c defines some hypersurface, say $\mathscr{S}_i$. The optimum should be in all these hypersurfaces simultaneously, i.e., $\mathbf{v}^*$ should be in some hypersurface $G \triangleq \cap_{i=1}^m \mathscr{S}_i$. As G is an hypersurface on its own, there exists a mapping $g : \mathbb{R}^m \to \mathbb{R}$ such that $G = \{\mathbf{v} | g(\mathbf{v}) = 0\}$. The smoothness of f_i^c implies that g is smooth too. Lagrange's theory states that at the optimum the gradients of J and g are parallel. That is $J_\mathbf{v}(\mathbf{v}^*) = \lambda g_\mathbf{v}(\mathbf{v}^*)$ for some scalar λ. On the one hand, the gradient $g_\mathbf{v}$ is perpendicular to G. On the other hand, each gradient $f_{i\mathbf{v}}^c$ is also perpendicular to G, for any $i = 1, \ldots, m$ (as $G \subseteq \mathscr{S}_i$). Therefore $g_\mathbf{v}(\mathbf{v}^*)$ is in a plane that is spanned by $\{f_{i\mathbf{v}}^c(\mathbf{v}^*)\}^1$ and $g_\mathbf{v}(\mathbf{v}^*)$ can be presented as a linear combination $g_\mathbf{v}(\mathbf{v}^*) = -(\boldsymbol{\mu}^*)^T \mathbf{f}_\mathbf{v}^c(\mathbf{v}^*)$ [92]. This form is more convenient as it does not require an explicit formulation of g. It leads to the well known necessary condition:

$$-J_\mathbf{v}(\mathbf{v}^*) = \sum_{i=1}^m \mu_i^* f_{i\mathbf{v}}^c(\mathbf{v}^*)$$

The vector on the left hand side is called the *steepest descent* and points outside G, i.e., in a direction that will improve J. The right hand side describes how each $f_{i\mathbf{v}}^c(\mathbf{v}^*)$ contributes to the steepest descent. If $\mu_i^* > 0$, then the direction of $f_{i\mathbf{v}}^c(\mathbf{v}^*)$ coincides with the steepest descent; If $\mu_i^* < 0$, then it is opposed to the steepest descent; and if $\mu_i^* = 0$ it does not contribute.

Let's go back to the original inequality constrained problem. Assume that $\mathbf{v}^*$ is an inequality constrained optimum and that $(\mu_i^*)_{i=1}^m$ are KKT multipliers. Theorem 3.3 implies that a $\mathbf{v}^*$ that satisfies $\{f_i^c(\mathbf{v}^*) \leq 0\}_{i=1}^m$ is required to satisfy conditions that are quite similar to those of a $\mathbf{v}^*$ that satisfy $\{f_i^c(\mathbf{v}^*) = 0\}_{i=1}^m$, only that: (a) not all the constraints are taken into account. Some constraints are considered *active* and some are *inactive*; (b) $(\mu_i^*)_{i=1}^m$ are nonnegative. An active f_i^c implies that J could gain better value if $f_i^c(\mathbf{v}^*) \leq 0$ was ignored, and therefore the i-th constraint is essential for the computation of $\mathbf{v}^*$. Otherwise, the i-th constraint is inactive and the same $\mathbf{v}^*$ can be found by ignoring it.

---

[1]It is assumed here that $\mathbf{v}^*$ is a regular point of the constraints, i.e., $(f_{i\mathbf{v}}^c(\mathbf{v}^*))_{i=1}^m$ are linearly independent. See p. 298 in [66].

An active constraint should satisfy the following conditions:

- $f_i^c(\mathbf{v}^*) = 0$. That means that $\mathbf{v}^*$ is located on the boundary of the corresponding admissible set $\mathcal{V}_i \triangleq \{\mathbf{v} | f_i^c(\mathbf{v}^*) \leq 0\}$.

- $\mu_i^* > 0$. As $f_{i\mathbf{v}}^c$ is aimed at a direction in which f_i^c increases, and as $f_i^c(\mathbf{v}^*) = 0$, a positive μ_i^* means that the steepest descent is pointing towards a region, where $f_i^c(\mathbf{v}^*) > 0$, which is an inadmissible region. In other words, a better $\mathbf{v}$ exists only in the inadmissible region, at least in the local sense.

If these conditions are not met then the constraint is inactive and $\mu_i^* = 0$. This concept is expressed by Eqs. (3.13a) and (3.13b).

Note that if $f_i^c(\mathbf{v}^*) = 0$ and $\mathbf{v}^*$ is an unconstrained optimum, then the steepest descent does not point toward the inadmissible region, therefore the constraint is inactive. This idea will be used later in Lemma 6.3.

The interpretation described above is illustrated by Fig. 3.1. In this figure, the constraints are described by three cylinders whose axes are perpendicular to the figure's plane. The contour lines describe the values of J. $\hat{\mathbf{v}}$ is an unconstrained optimum and $\mathcal{V}_i$ is the set of points that satisfy the constraint defined by f_i^c. It can be seen that f_1^c and f_2^c are inactive, and that $f_{3\mathbf{v}}^c(\mathbf{v}^*)$ points toward $-J_\mathbf{v}(\mathbf{v}^*)$.

So far for static optimization. When addressing an inequality constrained control optimization problem, it can be considered as a static one with an infinite number of optimization parameters, which should satisfy a state equation.

When the performance index form corresponds to Eq. (3.4), the problem is defined by

$$\dot{\mathbf{x}}(t) = \mathbf{f}(t, \mathbf{x}(t), \mathbf{u}(t)); \quad \mathbf{x}(0), \forall t \in (0, t_f) \tag{3.14a}$$

$$\{f_i^c(t, \mathbf{x}(t), \mathbf{u}(t)) \leq 0\}_{i=1}^m \tag{3.14b}$$

$$J(\mathbf{x}, \mathbf{u}) = l_f(\mathbf{x}(t_f)) + \int_0^{t_f} l(t, \mathbf{x}(t), \mathbf{u}(t)) \mathrm{d}t \tag{3.14c}$$

$(\mathbf{x}, \mathbf{u})$ is an admissible process if it satisfies Eq. (3.14a) and $f_i^c(t, \mathbf{x}(t), \mathbf{u}(t)) \leq 0$ for all $i \in \{1, \ldots, m\}$ and $t \in [0, t_f]$.

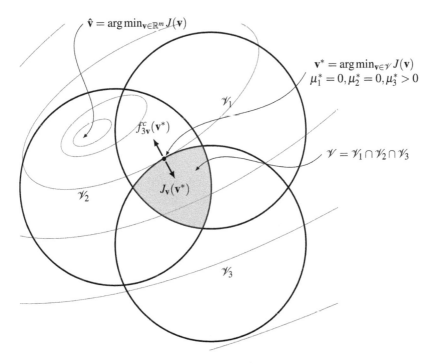

Figure 3.1: An illustration of an inequality constrained problem.

Let $\mathbf{f}^c : \mathbb{R} \times \mathbb{R}^n \times \mathbb{R}^{n_u} \to \mathbb{R}^m$, and $\boldsymbol{\mu} : \mathbb{R} \to \mathbb{R}^m$ be mappings of the inequality constraints and the KKT multipliers trajectory, respectively. By augmenting the performance index, a corresponding Hamiltonian function can be formulated:

$$J^a(\mathbf{x},\mathbf{u},\mathbf{p},\boldsymbol{\mu}) = l_f(\mathbf{x}(t_f)) + \int\limits_0^{t_f} \left(H(t,\mathbf{x}(t),\mathbf{u}(t),\mathbf{p}(t),\boldsymbol{\mu}(t)) - \mathbf{p}(t)^T \dot{\mathbf{x}}(t) \right) \mathrm{d}t$$

$$H(t,\mathbf{x}(t),\mathbf{u}(t),\mathbf{p}(t),\boldsymbol{\mu}(t)) \triangleq l(t,\mathbf{x}(t),\mathbf{u}(t)) + \mathbf{p}(t)^T \mathbf{f}(t,\mathbf{x}(t),\mathbf{u}(t)) \\ + \boldsymbol{\mu}(t)^T \mathbf{f}^c(t,\mathbf{x}(t),\mathbf{u}(t))$$

and according to the KKT theorem, at the optimum it is necessary that [16]:

$$J^a_{(\mathbf{x},\mathbf{u},\mathbf{p})}(\mathbf{x}^*,\mathbf{u}^*,\mathbf{p}^*,\boldsymbol{\mu}^*)^T = \mathbf{0}; \qquad \mu_i^*(t) \begin{cases} \geq 0 & , f_i^c(t,\mathbf{x}(t),\mathbf{u}(t)) = 0 \\ = 0 & , f_i^c(t,\mathbf{x}(t),\mathbf{u}(t)) < 0 \end{cases}$$

These conditions are summarized in the following theorem.

Theorem 3.4 [16]
Let an inequality constrained control optimization problem be defined by the following state equation, inequality constraints and performance index:

$$\dot{\mathbf{x}}(t) = \mathbf{f}(t, \mathbf{x}(t), \mathbf{u}(t)); \quad \mathbf{x}(0), \forall t \in (0, t_f)$$
$$\{f_i^c(t, \mathbf{x}(t), \mathbf{u}(t)) \le 0\}_{i=1}^m$$

$$J(\mathbf{x}, \mathbf{u}) = l_f(\mathbf{x}(t_f)) + \int_0^{t_f} l(t, \mathbf{x}(t), \mathbf{u}(t)) \mathrm{d}t$$

If $(\mathbf{x}^*, \mathbf{u}^*)$ *is an admissible optimum then it satisfies the following statements:*

$$\dot{\mathbf{x}}^*(t) = \mathbf{f}(t, \mathbf{x}^*(t), \mathbf{u}^*(t)); \quad \mathbf{x}^*(0) \tag{3.15a}$$

$$\dot{\mathbf{p}}^*(t) = -\mathbf{f}_{\mathbf{x}}^T(t, \mathbf{x}^*(t), \mathbf{u}^*(t)) \mathbf{p}^*(t) - l_{\mathbf{x}}(t, \mathbf{x}^*(t), \mathbf{u}^*(t))^T$$
$$- \sum_{i=1}^m f_{i,\mathbf{x}}^c(t, \mathbf{x}^*(t), \mathbf{u}^*(t))^T \mu_i^*(t); \quad \mathbf{p}^*(t_f) = (l_f)_{\mathbf{x}}(\mathbf{x}(t_f))^T \tag{3.15b}$$

$$-\mathbf{f}_{\mathbf{u}}^T(t, \mathbf{x}^*(t), \mathbf{u}^*(t)) \mathbf{p}^*(t) - l_{\mathbf{u}}(t, \mathbf{x}^*(t), \mathbf{u}^*(t))^T$$
$$- \sum_{i=1}^m f_{i,\mathbf{u}}^c(t, \mathbf{x}^*(t), \mathbf{u}^*(t))^T \mu_i^*(t) = \mathbf{0} \tag{3.15c}$$

and

$$\mu_i^*(t) \begin{cases} \ge 0 & , f_i^c(t, \mathbf{x}^*(t), \mathbf{u}^*(t)) = 0 \\ = 0 & , f_i^c(t, \mathbf{x}^*(t), \mathbf{u}^*(t)) < 0 \end{cases}$$

for all $t \in (0, t_f)$.

It should be noted that most of the time, inequality constrained optimization leads to nonlinear control law. Therefore, when inequality constrained optimal design is being carried out, the closed loop system will be nonlinear, even for LTI plants.

3.4 Krotov's Sufficient Conditions

In many optimal control problems it is customary to formulate suitable solution methods by either first order variational calculus or Pontryagin's minimum principle. However, since for many problems these theorems provide necessary conditions [53], the calculated solution is merely a candidate optimum but not a guaranteed one. Additional steps should be carried out to make sure that the derived solution is indeed an optimum. Furthermore, even if the solution was found to be an optimum, it might be local rather than a global one.

In the sixties of the previous century, series of results on sufficient conditions for global optimum of optimal control problems began to be published by V. F. Krotov [55]. A brief description of the continuous version of this approach with main theorems is given hereinafter. The theorems are taken from the published works of Krotov but the formulation was adapted to the needs and nature of the discussed problems. A more comprehensive and general description of these results can be found in Krotov's works [56, 55].

Let Π be a set of smooth functions $\mathbb{R} \times \mathbb{R}^n \to \mathbb{R}$. The following theorem states that each $q \in \Pi$ is related to some equivalent formulation of the performance index.

Theorem 3.5
Let a constrained control optimization problem be defined by the state equation, performance index and the set of admissible control trajectories:

$$\dot{\mathbf{x}}(t) = \mathbf{f}(t, \mathbf{x}(t), \mathbf{u}(t)); \quad \mathbf{x}(0), \forall t \in (0, t_f) \tag{3.16a}$$

$$J(\mathbf{x}, \mathbf{u}) = l_f(\mathbf{x}(t_f)) + \int_0^{t_f} l(t, \mathbf{x}(t), \mathbf{u}(t)) dt \tag{3.16b}$$

$$\mathbf{u} \in \mathcal{U}$$

Let q be a given function in Π and let $(\mathbf{x}, \mathbf{u})$ be an admissible process. Then $J_{eq} : \mathcal{X} \times \mathcal{U} \to [0, \infty)$ is an equivalent representation of J that corresponds to q and is defined by:

$$J(\mathbf{x}, \mathbf{u}) = J_{eq}(\mathbf{x}, \mathbf{u}) = s_f(\mathbf{x}(t_f)) + q(0, \mathbf{x}(0)) + \int_0^{t_f} s(t, \mathbf{x}(t), \mathbf{u}(t)) dt \tag{3.17}$$

where

$$s(t, \boldsymbol{\xi}, \mathbf{v}) \triangleq q_t(t, \boldsymbol{\xi}) + q_{\mathbf{x}}(t, \boldsymbol{\xi}) \mathbf{f}(t, \boldsymbol{\xi}, \mathbf{v}) + l(t, \boldsymbol{\xi}, \mathbf{v}) \tag{3.18}$$

$$s_f(\boldsymbol{\xi}) \triangleq l_f(\boldsymbol{\xi}) - q(t_f, \boldsymbol{\xi}) \tag{3.19}$$

Here $\boldsymbol{\xi} \in \mathbb{R}^n$, $\mathbf{v} \in \mathbb{R}^{n_u}$.

Proof 3.1 Substituting Eqs. (3.16b) and (3.17)-(3.19) into $J_{eq} - J$:

$$J_{eq}(\mathbf{x}, \mathbf{u}) - J(\mathbf{x}, \mathbf{u}) = s_f(\mathbf{x}(t_f)) + q(0, \mathbf{x}(0)) - l_f(\mathbf{x}(t_f))$$

$$+ \int_0^{t_f} (s(t, \mathbf{x}(t), \mathbf{u}(t)) - l(t, \mathbf{x}(t), \mathbf{u}(t))) \, dt$$

$$= q(0, \mathbf{x}(0)) - q(t_f, \mathbf{x}(t_f))$$

$$+ \int_0^{t_f} (q_t(t, \mathbf{x}(t)) + q_{\mathbf{x}}(t, \mathbf{x}(t)) \mathbf{f}(t, \mathbf{x}(t), \mathbf{u}(t))) \, dt$$

As $(\mathbf{x}, \mathbf{u})$ is an admissible process:

$$J_{eq}(\mathbf{x}, \mathbf{u}) - J(\mathbf{x}, \mathbf{u}) = q(0, \mathbf{x}(0)) - q(t_f, \mathbf{x}(t_f))$$

$$+ \int_0^{t_f} (q_t(t, \mathbf{x}(t)) + q_{\mathbf{x}}(t, \mathbf{x}(t)) \dot{\mathbf{x}}(t)) \, dt$$

$$= q(0, \mathbf{x}(0)) - q(t_f, \mathbf{x}(t_f)) + \int_0^{t_f} dq(t, \mathbf{x}(t)) = 0$$

by the virtue of Newton-Leibnitz formula. Hence

$$J_{eq}(\mathbf{x}, \mathbf{u}) = J(\mathbf{x}, \mathbf{u})$$

for any given q.

The following theorem provides a sufficient condition for a global optimum $(\mathbf{x}^*, \mathbf{u}^*)$ by means of J_{eq}. The notation $\mathscr{X}(t)$ refers to the set $\{\mathbf{x}(t) | \mathbf{x} \in \mathscr{X}\} \subseteq \mathbb{R}^n$, i.e., an intersection of $\mathscr{X}$ at a given t. For instance, $\mathscr{X}(t_f) \subseteq \mathbb{R}^n$ is a set of all the terminal states of the state trajectories in $\mathscr{X}$.

Theorem 3.6
Let s and s_f be related to some q and let $(\mathbf{x}^*, \mathbf{u}^*)$ be an admissible process. If:

$$s(t, \mathbf{x}^*(t), \mathbf{u}^*(t)) = \min_{\substack{\boldsymbol{\xi} \in \mathscr{X}(t) \\ \mathbf{v} \in \mathscr{U}(t)}} s(t, \boldsymbol{\xi}, \mathbf{v}) \; \forall t \in [0, t_f) \qquad (3.20a)$$

$$s_f(\mathbf{x}^*(t_f)) = \min_{\boldsymbol{\xi} \in \mathscr{X}(t)} s_f(\boldsymbol{\xi}) \qquad (3.20b)$$

then $(\mathbf{x}^*, \mathbf{u}^*)$ is an optimal process.

Proof 3.2   Assume that Eqs. (3.20) holds. According to Theorem 3.5:

$$
\begin{aligned}
J(\mathbf{x},\mathbf{u}) - J(\mathbf{x}^*,\mathbf{u}^*) =& J_{eq}(\mathbf{x},\mathbf{u}) - J_{eq}(\mathbf{x}^*,\mathbf{u}^*) \\
=& s_f(\mathbf{x}(t_f)) - s_f(\mathbf{x}^*(t_f)) \\
& + \int_0^{t_f} (s(t,\mathbf{x}(t),\mathbf{u}(t)) - s(t,\mathbf{x}^*(t),\mathbf{u}^*(t))) \, dt \geq 0
\end{aligned}
$$

which assures that:

$$
J(\mathbf{x},\mathbf{u}) \geq J(\mathbf{x}^*,\mathbf{u}^*)
$$

for any admissible process $(\mathbf{x},\mathbf{u})$.

Remarks.

■ It is customary to denote q for which Eq. (3.20) holds as *Krotov function* or *solving function*. This solving function is not unique.

■ An optimum derived by this theorem is global since the minimization problem defined in Eq. (3.20) is global.

■ Note that the equivalence $J(\mathbf{x},\mathbf{u}) = J_{eq}(\mathbf{x},\mathbf{u})$ holds if $\mathbf{x}$ and $\mathbf{u}$ satisfies the state equation (Eq. (3.16a)). Otherwise, Eq. (3.20a) might be non-zero.

■ As q is not unique, J_{eq}, s and s_f are non-unique too.

■ In the published works of Krotov (such as [55, 56, 57]), s and q are defined with an opposite signs which turn some of the minimization problems into maximization ones. In our research we use a formulation that leaves the problem as the customary one, i.e., a minimization problem. Though, there is no essential difference between these two formulations.

■ The formulations above are defined with an assumption that the functions in Π are smooth. This assumption can be weakened into smoothness over $\mathscr{X}$; and continuity and piecewise differentiability over $(0,t_f)$ [56]. That is, $q \in \Pi$ should be differentiable at any $t \in (0,t_f)$ except for some set of t's with a zero measure.

Besides being useful for defining a sufficient condition for a global optimum, the equivalent representation of the performance index forms a basis for different algorithms, aimed at solving optimal control problems [55]. One of these algorithms is Krotov's method, which will be described in Section 4.3.

Chapter 4

Optimal Control: Successive Solution Methods

This chapter overviews three known successive methods that can be used for computing an approximate solution to optimal control problems. These methods are described here generally and additional derivations, suitable to the given problem, should be carried out before they can be applied.

4.1 Steepest Descent

Steepest descent (a.k.a *gradient descent*) method, is a simple and well known tool, that sometimes can be used for solving optimal control problems numerically [72, 15]. Let a finite set of control parameters be $\tilde{\mathbf{u}}$ and let J be a smooth performance index. Minimization of J is carried out successively by finding a sequence $\{\tilde{\mathbf{u}}_k\}$ that converges to the candidate optimum - $\tilde{\mathbf{u}}^*$. At each step, the algorithm propagates from $\tilde{\mathbf{u}}_k$ to $\tilde{\mathbf{u}}_{k+1}$ by performing a small variation in $\mathbf{u}$, on the one hand, which causes the sharpest negative variation in J's, on the other hand.

Let a $\varepsilon\boldsymbol{\delta}$ be some variation of $\tilde{\mathbf{u}}_k$ such that $\boldsymbol{\delta} \in \mathbb{R}^{n_u}$ and $\varepsilon > 0$. By simple calculus:

$$J(\tilde{\mathbf{u}}_k + \varepsilon\boldsymbol{\delta}) - J(\tilde{\mathbf{u}}_k) = \int_{J(\tilde{\mathbf{u}}_k)}^{J(\tilde{\mathbf{u}}_k + \varepsilon\boldsymbol{\delta})} \mathrm{d}J(\mathbf{u}) = \sum_{i=1}^{n_u} \int_{\tilde{u}_{k,i}}^{\tilde{u}_{k,i} + \varepsilon\delta_i} J_{\tilde{u}_i}(\mathbf{u})\mathrm{d}u_i$$

$$= \int_0^\varepsilon J_{\tilde{\mathbf{u}}}(\tilde{\mathbf{u}}_k + x\boldsymbol{\delta})\boldsymbol{\delta}\,\mathrm{d}x$$

where $J_{\tilde{\mathbf{u}}} : \mathbb{R}^{n_u} \to \mathbb{R}^{n_u}$ is the gradient of J. Here $J_{\tilde{\mathbf{u}}}(\tilde{\mathbf{u}})$ is a row vector. Let $\boldsymbol{\delta} = -J_{\tilde{\mathbf{u}}}(\tilde{\mathbf{u}}_k)^T$:

$$J(\tilde{\mathbf{u}}_k - \varepsilon J_{\tilde{\mathbf{u}}}(\tilde{\mathbf{u}}_k)^T) - J(\tilde{\mathbf{u}}_k) = -\int_0^\varepsilon J_{\tilde{\mathbf{u}}}(\tilde{\mathbf{u}}_k - x J_{\tilde{\mathbf{u}}}(\tilde{\mathbf{u}}_k)^T)J_{\tilde{\mathbf{u}}}(\tilde{\mathbf{u}}_k)^T\mathrm{d}x \qquad (4.1)$$

J's smoothness implies that $J_{\mathbf{u}}$ is continuous. Hence $f(x) \triangleq J_{\tilde{\mathbf{u}}}(\tilde{\mathbf{u}}_k - xJ_{\tilde{\mathbf{u}}}(\tilde{\mathbf{u}}_k)^T)J_{\tilde{\mathbf{u}}}(\tilde{\mathbf{u}}_k)^T$ is continuous. Note that $f(0) \geq 0$. f's continuity says that for any $r \geq 0$ there exists a corresponding $\varepsilon \geq 0$ such that for any $x \in [0, \varepsilon]$ we have $f(x) \in [f(0) - r, f(0) + r]$. By letting $r = f(0)$, there exists $\varepsilon > 0$ such that $f(x) \in [0, 2f(0)]$ if $x \in [0, \varepsilon]$. Hence, there exists ε such that $J_{\tilde{\mathbf{u}}}(\tilde{\mathbf{u}}_k - xJ_{\tilde{\mathbf{u}}}(\tilde{\mathbf{u}}_k)^T)J_{\tilde{\mathbf{u}}}(\tilde{\mathbf{u}}_k)^T \geq 0$ for all $x \in [0, \varepsilon]$. By Eq. (4.1):

$$J(\tilde{\mathbf{u}}_k - \varepsilon J_{\tilde{\mathbf{u}}}(\tilde{\mathbf{u}}_k)^T) - J(\tilde{\mathbf{u}}_k) \leq 0$$

The steepest decent method takes advantage of this property in order to compute a better $\tilde{\mathbf{u}}_{k+1}$ from a given $\tilde{\mathbf{u}}_k$ by

$$\tilde{\mathbf{u}}_{k+1} = \tilde{\mathbf{u}}_k - \varepsilon J_{\tilde{\mathbf{u}}}^T(\tilde{\mathbf{u}}_k)$$

where ε is small enough to assure improvement. However the method's utilization depends on the differentiability of J in each $\tilde{\mathbf{u}}_k$ and the convergence tends to be quite slow due to its restriction to a small ε.

The name *steepest descent* rises from the fact that by letting $\boldsymbol{\delta} = -J_{\tilde{\mathbf{u}}}(\tilde{\mathbf{u}}_k)^T$ it is not only that $f(0) \geq 0$, but it also satisfies $f(0) \geq J_{\tilde{\mathbf{u}}}(\tilde{\mathbf{u}}_k)\mathbf{v}/\|\mathbf{v}\|$ for any $\mathbf{v} \in \mathbb{R}^{n_u}$. It follows from the fundamental property of linear functionals [25]:

$$J_{\tilde{\mathbf{u}}}(\tilde{\mathbf{u}}_k)^T = \arg\sup_{\mathbf{v}} \left\{ J_{\tilde{\mathbf{u}}}(\tilde{\mathbf{u}}_k)\mathbf{v}\frac{1}{\|\mathbf{v}\|} \right\}$$

This approach can be extended from problems with a finite set of control parameters to computation of optimal control trajectories that satisfy Eqs. (3.9a-c) [52, 15].

The required steps are summarized in the following algorithm. It initializes with choosing some initial control trajectory - $\mathbf{u}_0$ and then iterating for $k = \{0, 1, 2, \ldots\}$. In each iteration:

1. Solve:

$$\dot{\mathbf{x}}_k(t) = \mathbf{f}(t, \mathbf{x}_k(t), \mathbf{u}_k(t)); \quad \mathbf{x}_k(0) = \mathbf{x}(0)$$
$$\dot{\mathbf{p}}_k(t) = -\mathbf{f_x}(t, \mathbf{x}_k(t), \mathbf{u}_k(t))^T \mathbf{p}_k(t) - l_\mathbf{x}(t, \mathbf{x}_k(t), \mathbf{u}_k(t))^T$$
$$\mathbf{p}(t_f) = l_{f\mathbf{x}}(\mathbf{x}_k(t_f))^T$$

2. Set:

$$H_\mathbf{u}(t) \triangleq -\mathbf{p}_k(t)^T \mathbf{f_u}(t, \mathbf{x}_k(t), \mathbf{u}_k(t)) - l_\mathbf{u}(t, \mathbf{x}_k(t), \mathbf{u}_k(t))$$

3. Propagate to the improved control trajectory:

$$\mathbf{u}_{k+1}(t) = \mathbf{u}_k(t) - \varepsilon H_\mathbf{u}(t)$$

where $\varepsilon > 0$ is some small scalar.

Remarks.

■ Essentially, numerical implementation of this method is done with respect to some set of time samples $\{t_i\}$ rather than the continuous time axis.

■ The iterations stop when $\|H_\mathbf{u}(t)\|$ becomes small enough.

4.2 Parametric Optimization: Newton's Method

In optimization, Newton's method is an iterative method for computing stationary solutions. Let $\tilde{\mathbf{u}}$ be a vector of control parameters. In this method the minimization of J is carried out successively by finding a sequence $\{\tilde{\mathbf{u}}_k\}$ that converges to a candidate optimum - $\tilde{\mathbf{u}}^*$. In each step, the algorithm computes the next point by performing a local minimization of J's 2nd order Taylor polynomial at the current point's neighborhood, as follows.

The Taylor expansion of J at the neighborhood of $\tilde{\mathbf{u}}$ is:

$$J(\tilde{\mathbf{u}} + \boldsymbol{\delta}) = J(\tilde{\mathbf{u}}) + J_{\tilde{\mathbf{u}}}(\tilde{\mathbf{u}})\boldsymbol{\delta} + 0.5\boldsymbol{\delta}^T J_{\tilde{\mathbf{u}}\tilde{\mathbf{u}}}(\tilde{\mathbf{u}})\boldsymbol{\delta} + o_2(\|\boldsymbol{\delta}\|)$$
$$= f(\boldsymbol{\delta}) + o_2(\|\boldsymbol{\delta}\|)$$
$$f(\boldsymbol{\delta}) \triangleq J(\tilde{\mathbf{u}}) + J_{\tilde{\mathbf{u}}}(\tilde{\mathbf{u}})\boldsymbol{\delta} + 0.5\boldsymbol{\delta}^T J_{\tilde{\mathbf{u}}\tilde{\mathbf{u}}}(\tilde{\mathbf{u}})\boldsymbol{\delta}$$

where $\boldsymbol{\delta}$ is some small perturbation; $J_{\tilde{\mathbf{u}}}(\tilde{\mathbf{u}})$ and $J_{\tilde{\mathbf{u}}\tilde{\mathbf{u}}}(\tilde{\mathbf{u}})$ are J's gradient vector and Hessian matrix at $\tilde{\mathbf{u}}$, respectively; f is the 2nd order Taylor polynomial and o_2 is the remainder of f. The goal is to determine $\boldsymbol{\delta}$ that minimizes $f(\boldsymbol{\delta})$. If $J_{\tilde{\mathbf{u}}\tilde{\mathbf{u}}}(\tilde{\mathbf{u}}) > 0$ then:

$$f(\boldsymbol{\delta}) = J(\tilde{\mathbf{u}}) + \frac{1}{2}\left(J_{\tilde{\mathbf{u}}}(\tilde{\mathbf{u}})^T + J_{\tilde{\mathbf{u}}\tilde{\mathbf{u}}}(\tilde{\mathbf{u}})\boldsymbol{\delta}\right)^T J_{\tilde{\mathbf{u}}\tilde{\mathbf{u}}}(\tilde{\mathbf{u}})^{-1}\left(J_{\tilde{\mathbf{u}}}(\tilde{\mathbf{u}})^T + J_{\tilde{\mathbf{u}}\tilde{\mathbf{u}}}(\tilde{\mathbf{u}})\boldsymbol{\delta}\right)$$
$$- \frac{1}{2}J_{\tilde{\mathbf{u}}}(\tilde{\mathbf{u}})J_{\tilde{\mathbf{u}}\tilde{\mathbf{u}}}(\tilde{\mathbf{u}})^{-1}J_{\tilde{\mathbf{u}}}(\tilde{\mathbf{u}})^T$$

It is evident that

$$\arg\min_{\boldsymbol{\delta}} f(\boldsymbol{\delta}) = -J_{\tilde{\mathbf{u}}\tilde{\mathbf{u}}}(\tilde{\mathbf{u}})^{-1}J_{\tilde{\mathbf{u}}}(\tilde{\mathbf{u}})^T$$

Computing $\tilde{\mathbf{u}}^{k+1}$ from a given $\tilde{\mathbf{u}}^k$ is carried out by using this approach sequentially and ignoring the term o_2 even for large perturbations. That is:

$$\tilde{\mathbf{u}}^{k+1} = \tilde{\mathbf{u}}^k - J_{\tilde{\mathbf{u}}\tilde{\mathbf{u}}}^{-1}(\tilde{\mathbf{u}}^k)J_{\tilde{\mathbf{u}}}^T(\tilde{\mathbf{u}}^k)$$

Essentially, the method requires to compute the gradient vector and Hessian matrix in each point in the sequence [66]. The method's effectiveness depends on the differentiability of J and the non-negativeness of $J_{\tilde{\mathbf{u}}\mathbf{u}}(\tilde{\mathbf{u}}^k)$ in each $\tilde{\mathbf{u}}^k$ in the sequence.

4.3 Krotov's Method—Successive Global Improvements of Control

The purpose of Krotov's method is to solve optimal control problems numerically by taking advantage of theoretical concepts from Section 3.4. According to this method, the key to the problem is formulating a sequence of functions with special properties. If such sequence can be found, it allows computing a process that is a candidate global optimum.

The method was used successfully for solving optimal control problems in quantum mechanics [57] and was recommended for solving an optimal control problem, defined by linear differential equations with controlled coefficients. Hamilton systems characterized by compact representability in the complex form and availability of dynamic invariant, were considered [57].

Before introducing this method, an *improving sequence* is defined.

Definition 4.1   Let $\{(\mathbf{x}_k, \mathbf{u}_k)\}$ be a sequence of admissible processes and assume that $\inf_{\substack{\mathbf{x} \in \mathscr{X} \\ \mathbf{u} \in \mathscr{U}}} J(\mathbf{x}, \mathbf{u})$ exists. If

$$J(\mathbf{x}_k, \mathbf{u}_k) \geq J(\mathbf{x}_{k+1}, \mathbf{u}_{k+1}) \tag{4.2}$$

for all $k = 1, 2, \ldots$ and:

$$\lim_{k \to \infty} J(\mathbf{x}_k, \mathbf{u}_k) \tag{4.3}$$

exists, then $\{(\mathbf{x}_k, \mathbf{u}_k)\}$ is called an improving sequence.

If an improving sequence can be found, it allows computing an admissible process that is 'arbitrary close' to one that satisfies Pontryagin's minimum principle. Krotov's method is aimed at computing such sequence by successively improving admissible processes. The concept that underlies these improvements is a sufficient condition for process improvement, given in the following theorem. Recall that each $q \in \Pi$ is related to some s and s_f, defined in Eqs. (3.18) and (3.19). Here s_k and $s_{f,k}$ signifies s and s_f that corresponds a function q_k.

Theorem 4.1
Let a given admissible process be $(\mathbf{x}_k, \mathbf{u}_k)$ and let $q_k \in \Pi$. If $(\mathbf{x}_k, \mathbf{u}_k)$ is a solution to the following maximization problem:

$$\begin{aligned} s_k(t, \mathbf{x}_k(t), \mathbf{u}_k(t)) &= \max_{\boldsymbol{\xi} \in \mathscr{X}(t)} s_k(t, \boldsymbol{\xi}, \mathbf{u}_k(t)) \\ s_{f,k}(\mathbf{x}_k(t_f)) &= \max_{\boldsymbol{\xi} \in \mathscr{X}(t_f)} s_{f,k}(\boldsymbol{\xi}) \end{aligned} \tag{4.4}$$

and if $\hat{\mathbf{u}}_{k+1}$ is a feedback that satisfies

$$\hat{\mathbf{u}}_{k+1}(t, \boldsymbol{\xi}) = \arg \min_{\mathbf{v} \in \mathscr{U}(t)} s_k(t, \boldsymbol{\xi}, \mathbf{v}); \quad \forall t \in [0, t_f] \tag{4.5}$$

then $\mathbf{x}_{k+1}$ that solves:

$$\dot{\mathbf{x}}_{k+1}(t) = \mathbf{f}(t, \mathbf{x}_{k+1}(t), \hat{\mathbf{u}}_{k+1}(t, \mathbf{x}_{k+1}(t))); \quad \mathbf{x}_{k+1}(0) = \mathbf{x}(0), \forall t \in (0, t_f) \tag{4.6}$$

and the control trajectory $\mathbf{u}_{k+1}(t) = \hat{\mathbf{u}}_{k+1}(t, \mathbf{x}_{k+1}(t))$, satisfy Eq. (4.2).

Proof 4.1   Assume that Eqs. (4.4) and (4.5) hold. Then

$$J(\mathbf{x}_k,\mathbf{u}_k) - J(\mathbf{x}_{k+1},\mathbf{u}_{k+1}) = J_{eq}(\mathbf{x}_k,\mathbf{u}_k) - J_{eq}(\mathbf{x}_{k+1},\mathbf{u}_{k+1}) \tag{4.7}$$

$$= s_{f,k}(\mathbf{x}_k(t_f)) - s_{f,k}(\mathbf{x}_{k+1}(t_f)) + \int_0^{t_f} (s_k(t,\mathbf{x}_k,\mathbf{u}_k) - s_k(t,\mathbf{x}_{k+1},\mathbf{u}_{k+1}))\,dt$$

$$= s_{f,k}(\mathbf{x}_k(t_f)) - s_{f,k}(\mathbf{x}_{k+1}(t_f)) + \int_0^{t_f} (s_k(t,\mathbf{x}_k(t),\mathbf{u}_k(t)) - s_k(t,\mathbf{x}_{k+1}(t),\mathbf{u}_k(t)))\,dt$$

$$+ \int_0^{t_f} (s_k(t,\mathbf{x}_{k+1}(t),\mathbf{u}_k(t)) - s_k(t,\mathbf{x}_{k+1}(t),\mathbf{u}_{k+1}(t)))\,dt$$

Equation 4.4 infers that

$$s_{f,k}(\mathbf{x}_k(t_f)) - s_{f,k}(\mathbf{x}_{k+1}(t_f)) \geq 0; \quad s_k(t,\mathbf{x}_k(t),\mathbf{u}_k(t)) - s_k(t,\mathbf{x}_{k+1}(t),\mathbf{u}_k(t)) \geq 0$$

and from Eq. (4.5) follows that:

$$s_k(t,\mathbf{x}_{k+1}(t),\mathbf{u}_k(t)) - s_k(t,\mathbf{x}_{k+1}(t),\mathbf{u}_{k+1}(t)) \geq 0$$

Therefore $J(\mathbf{x}_k,\mathbf{u}_k) - J(\mathbf{x}_{k+1},\mathbf{u}_{k+1})$ is non negative.

Thus, if for a given $(\mathbf{x}_k,\mathbf{u}_k)$ one can find q_k such that Eq. (4.4) holds, then it is possible to compute an improved process $(\mathbf{x}_{k+1},\mathbf{u}_{k+1})$. Such q_k is called an *improving function*. By solving this problem successively an improving sequence is constructed and a candidate optimum is obtained.

The method is summarized in the following algorithm. The algorithm initialization requires a computation of some initial admissible process $(\mathbf{x}_0,\mathbf{u}_0)$. Afterwards, the following steps are iterated for $k = \{0,1,2,\ldots\}$ until convergence:

1. Find q_k that solves

$$s_k(t,\mathbf{x}_k(t),\mathbf{u}_k(t)) = \max_{\boldsymbol{\xi} \in \mathscr{X}(t)} s_k(t,\boldsymbol{\xi},\mathbf{u}_k(t))$$

$$s_{f,k}(\mathbf{x}_k(t_f)) = \max_{\boldsymbol{\xi} \in \mathscr{X}(t_f)} s_{f,k}(\boldsymbol{\xi})$$

   for a given $(\mathbf{x}_k,\mathbf{u}_k)$ and for all t in $[0,t_f)$.

2. Find a feedback $\hat{\mathbf{u}}_{k+1}$ such that

$$\hat{\mathbf{u}}_{k+1}(t,\mathbf{x}(t)) = \arg \min_{\mathbf{v} \in \mathscr{U}(t)} s_k(t,\mathbf{x}(t),\mathbf{v})$$

   for all t in $[0,t_f]$

3. Compute the next, improved state and control processes, by solving

$$\dot{\mathbf{x}}_{k+1}(t) = \mathbf{f}\big(t, \mathbf{x}_{k+1}(t), \hat{\mathbf{u}}_{k+1}(t, \mathbf{x}_{k+1}(t))\big)$$

and setting:

$$\mathbf{u}_{k+1}(t) = \hat{\mathbf{u}}_{k+1}(t, \mathbf{x}_{k+1}(t))$$

Remarks.

■ It was proved by Krotov [56] that this algorithm generates an improving sequence of processes $\{(\mathbf{x}_k, \mathbf{u}_k)\}$ such that $J(\mathbf{x}_k, \mathbf{u}_k) \geq J(\mathbf{x}_{k+1}, \mathbf{u}_{k+1})$.

■ The method has a significant advantage over algorithms that are based on small variations. The latter are constrained to small process variations, which is troublesome as: (1) it leads to slow convergence rate and (2) for some optimal control problems small variations are impossible [55].

■ Like in Lyapunov's stability method, the use of Krotov's method is not straightforward. It requires formulating a suitable sequence of improving functions $\{q_k\}$. These functions usually differ from one optimal control problem to another and therefore their formulation is not trivial. Even though, as of this writing, there is no known unified approach for exact formulation of these functions, a possible approach for their formulation was suggested [55]. It requires some parameters tuning by trial and error in each iteration until an improvement is obtained.

■ The essential non-uniqueness of the improving function is a key characteristic of this approach. This vagueness is an advantage and a disadvantage at the same time. On the one hand, it provides a solution method with high flexibility level but on the other hand, it poses an additional challenge to the control design.

■ Sometimes the improving function implies on the form of a corresponding Lyapunov function. When dealing with non-linear plants this is an important contribution, since for such systems stability is always questionable.

Chapter 5

Control Using Viscous Dampers

The properties of a viscous fluid damper (VD), whose force is described by Eq. (2.10), are governed by a single parameter—the viscous gain. Choosing this gain is one of the fundamental tasks in VD control design. When multiple gains are sought, which is a typical case in structural control, this task might turn quite challenging. Several approaches for competing with this problem can be found in the literature [4, 76].

The use of optimization methods for design is preferred by many engineers. That applies also to VD gains computation, thereby leading to optimal gains design problem. This problem is a parametric optimal control problem. It turns out that, in contrast to their practical simplicity, VDs optimal gains' computation is a quite hard problem. Its complete solution is not known and its nonlinear programming complexity was proved for the general case [6]. Corresponding optimal control problems were formulated by means of $\|H\|_2$ and $\|H\|_\infty$ system norms [107, 106, 93] and solution methods such as linear matrix inequalities (LMI) [107]; gradient based methods [4, 111, 93]; second-order [93] and quasi-newton methods [111], were suggested.

In this chapter, a method aimed at computing optimal viscous gains is presented. It addresses two cases. Section 5.1 deals with a free vibration case and Section 5.2 is focused on an externally excited one. The optimization objective is defined by a modified performance index that consists of a

$\|H\|_2$ system norm and a quadratic gains norm. An algorithm for computing a candidate optimum is provided. It is based on Newton's optimization method with an effective calculation method for the Hessian.

5.1 Optimal Control by Viscous Dampers: Free Vibration

Denote the viscous gain from Eq. (2.10) by $\tilde{u}$. The state-space equation of a plant with n_u VDs is:

$$\dot{\mathbf{x}}(t) = \mathbf{A}\mathbf{x}(t) - \sum_{i=1}^{n_u} \mathbf{b}_i \tilde{u}_i \mathbf{c}_i \mathbf{x}(t) \tag{5.1}$$

where every $\mathbf{c}_i^T \in \mathbb{R}^n$ satisfies

$$\mathbf{c}_i \mathbf{x}(t) = \dot{z}_{di}(t) \tag{5.2}$$

If the plant is modelled by Eq. (2.4), then $\mathbf{c}_i = \begin{bmatrix} \mathbf{0} & \boldsymbol{\psi}_i^T \end{bmatrix}$ where $\boldsymbol{\psi}_i$ is the i-th column of $\boldsymbol{\Psi}$ (see the first remark to Lemma 2.1).

Problem 5.1 (G) *[40, 45] Let an optimal control problem be defined by:*

$$\dot{\mathbf{x}}(t) = \left(\mathbf{A} - \sum_{i=1}^{n_u} \mathbf{b}_i \tilde{u}_i \mathbf{c}_i \right) \mathbf{x}(t); \quad \mathbf{x}(0), t \in (0, \infty) \tag{5.3a}$$

$$J(\mathbf{x}, \tilde{\mathbf{u}}) = \frac{1}{2} \int_0^{\infty} \mathbf{x}(t)^T \mathbf{Q}\mathbf{x}(t) \mathrm{d}t + \frac{1}{2} \tilde{\mathbf{u}}^T \mathbf{R}\tilde{\mathbf{u}} \tag{5.3b}$$

where $\tilde{\mathbf{u}} = (\tilde{u}_i)_{i=1}^{n_u} \in \mathbb{R}^{n_u}$ is a gains vector; $\mathbf{Q} \in \mathbb{R}^{n \times n}$ and $\mathbf{R} \in \mathbb{R}^{n_u \times n_u}$ are nonnegative weighting matrices.

For stable plants, J can be written in an algebraic form, as follows.

Lemma 5.1
If $\mathbf{A}_{cl}(\tilde{\mathbf{u}}) \triangleq \mathbf{A} - \sum_{i=1}^{n_u} \mathbf{b}_i \tilde{u}_i \mathbf{c}_i$ is Hurwitz, then:

$$J(\mathbf{x}, \tilde{\mathbf{u}}) = J(\mathbf{x}(0), \tilde{\mathbf{u}}) = \frac{1}{2} e(\mathbf{x}(0), \tilde{\mathbf{u}}) + \frac{1}{2} \tilde{\mathbf{u}}^T \mathbf{R}\tilde{\mathbf{u}} \tag{5.4}$$

where:

$$e(\mathbf{x}(0), \tilde{\mathbf{u}}) \triangleq \mathbf{x}^T(0) \mathbf{P}(\tilde{\mathbf{u}}) \mathbf{x}(0) \tag{5.5}$$

and $\mathbf{P} : \mathbb{R}^{n_u} \to \mathbb{R}^{n \times n}$ *satisfies the algebraic Lyapunov equation:*

$$\mathbf{P}(\tilde{\mathbf{u}})\mathbf{A}_{cl}(\tilde{\mathbf{u}}) + \mathbf{A}_{cl}(\tilde{\mathbf{u}})^T \mathbf{P}(\tilde{\mathbf{u}}) = -\mathbf{Q} \tag{5.6}$$

Proof 5.1   Since Eq. (5.3a) is LTI, the closed loop is

$$\mathbf{x}(t) = e^{\mathbf{A}_{cl}(\tilde{\mathbf{u}})t}\mathbf{x}(0)$$

Substitution into Eq. (5.3b) yields:

$$J(\mathbf{x}, \tilde{\mathbf{u}}) = \frac{1}{2}\mathbf{x}^T(0)\left(\int\limits_0^\infty e^{\mathbf{A}_{cl}(\tilde{\mathbf{u}})^T t}\mathbf{Q}(\tilde{\mathbf{u}})e^{\mathbf{A}_{cl}(\tilde{\mathbf{u}})t}\mathrm{d}t\right)\mathbf{x}(0) + 0.5\tilde{\mathbf{u}}^T\mathbf{R}\tilde{\mathbf{u}}$$

Since $\mathbf{A}_{cl}(\tilde{\mathbf{u}})$ is Hurwitz, the integral enclosed in the brackets can be calculated by solving 5.6 [96].

Remarks.

■ In the addressed problem $\mathbf{x}(0)$ is fixed and independent of the control gains. Therefore, its dependence will be omitted from the following derivations. Additionally, from brevity considerations $\mathbf{P}$ will be used instead of $\mathbf{P}(\tilde{\mathbf{u}})$.

■ Let the plant be excited by some input signal - $\ddot{z}_g : \mathbb{R} \to \mathbb{R}$ such that:

$$\dot{\mathbf{x}}(t) = \left(\mathbf{A} - \sum_{i=1}^{n_u}\mathbf{b}_i\tilde{u}_i\mathbf{c}_i\right)\mathbf{x}(t) + \mathbf{x}(0)\ddot{z}_g(t)$$

and let the initial condition be zero. It follows that $e(\tilde{\mathbf{u}})$ is the squared system norm $\|\mathbf{H}_{y\ddot{z}_g}(\tilde{\mathbf{u}})\|_2^2$ where $\mathbf{H}_{y\ddot{z}_g}$ is the transfer function matrix from $\ddot{z}_g$ to $\mathbf{y}(t) = \mathbf{Q}^{0.5}\mathbf{x}(t)$ and $\mathbf{Q}^{0.5}$ is the matrix square root of $\mathbf{Q}$. This point will be discussed in Section 5.2.

In order to use Newton's method for computing a candidate optimum, the gradient and the Hessian of e, with respect to $\tilde{\mathbf{u}}$, should be computed. Their closed-form computation is given by the following lemma and the subsequent theorem.

Lemma 5.2
Assume that $\mathbf{A}_{cl}(\tilde{\mathbf{u}})$ *is Hurwitz for a given* $\tilde{\mathbf{u}}$. *The gradient* $e_{\tilde{\mathbf{u}}}(\tilde{\mathbf{u}})$ *is a row vector constructed from the main diagonal of*

$$-2\mathbf{CLPB} \tag{5.7}$$

where $\mathbf{P}$, $\mathbf{L}$ *are the solutions of:*

$$\mathbf{P}\mathbf{A}_{cl}(\tilde{\mathbf{u}}) + \mathbf{A}_{cl}(\tilde{\mathbf{u}})^T\mathbf{P} = -\mathbf{Q} \tag{5.8a}$$

$$\mathbf{L}\mathbf{A}_{cl}^T(\tilde{\mathbf{u}}) + \mathbf{A}_{cl}(\tilde{\mathbf{u}})\mathbf{L} = -\mathbf{x}(0)\mathbf{x}(0)^T \tag{5.8b}$$

Proof 5.2   Recall that $\mathbf{P} : \mathbb{R}^{n_u} \to \mathbb{R}^{n \times n}$. Hence, the i-th element in the gradient of e is $e_{\tilde{u}_i}(\tilde{\mathbf{u}}) = \frac{\partial}{\partial \tilde{u}_i}(\mathbf{x}(0)^T\mathbf{P}\mathbf{x}(0)) = \mathbf{x}(0)^T\mathbf{P}_{\tilde{u}_i}\mathbf{x}(0)$. The full derivative of Eq. (5.8a) with respect to $\tilde{u}_i$ leads to:

$$\mathbf{P}_{\tilde{u}_i}\mathbf{A}_{cl}(\tilde{\mathbf{u}}) + \mathbf{A}_{cl}(\tilde{\mathbf{u}})^T\mathbf{P}_{\tilde{u}_i} = \mathbf{P}\mathbf{b}_i\mathbf{c}_i + \mathbf{c}_i^T\mathbf{b}_i^T\mathbf{P} \tag{5.9}$$

Let $y_1(t) \triangleq \mathbf{b}_i^T\mathbf{P}e^{\mathbf{A}_{cl}(\tilde{\mathbf{u}})t}\mathbf{x}(0)$ and $y_2(t) \triangleq \mathbf{c}_i e^{\mathbf{A}_{cl}(\tilde{\mathbf{u}})t}\mathbf{x}(0)$. Because $\mathbf{A}_{cl}(\tilde{\mathbf{u}})$ is Hurwitz, $\mathbf{P}_{\tilde{u}_i}$ can be written as a Lyapunov integral:

$$\mathbf{x}(0)^T\mathbf{P}_{\tilde{u}_i}\mathbf{x}(0) = -\int_0^\infty \mathbf{x}(0)^T e^{\mathbf{A}_{cl}^T(\tilde{\mathbf{u}})t}(\mathbf{P}\mathbf{b}_i\mathbf{c}_i + \mathbf{c}_i^T\mathbf{b}_i^T\mathbf{P})e^{\mathbf{A}_{cl}(\tilde{\mathbf{u}})t}\mathbf{x}(0)\mathrm{d}t$$

$$= -2\int_0^\infty y_2(t)y_1(t)\mathrm{d}t$$

On the other hand, $\mathbf{L}$ can also be written as Lyapunov integral:

$$\mathbf{L} = \int_0^\infty e^{\mathbf{A}_{cl}(\tilde{\mathbf{u}})t}\mathbf{x}(0)\mathbf{x}(0)^T e^{\mathbf{A}_{cl}^T(\tilde{\mathbf{u}})t}\mathrm{d}t$$

Hence, the i-th element in the main diagonal of Eq. (5.7) is

$$-2(\mathbf{CLPB})_{ii} = -2\mathbf{c}_i\mathbf{L}\mathbf{P}\mathbf{b}_i = -2\int_0^\infty \mathbf{c}_i e^{\mathbf{A}_{cl}(\tilde{\mathbf{u}})t}\mathbf{x}(0)\mathbf{x}(0)^T e^{\mathbf{A}_{cl}^T(\tilde{\mathbf{u}})t}\mathbf{P}\mathbf{b}_i\mathrm{d}t$$

$$= -2\int_0^\infty y_1(t)y_2(t)\mathrm{d}t$$

which is identical to $\mathbf{x}(0)^T\mathbf{P}_{\tilde{u}_i}\mathbf{x}(0)$.

Remarks.

- This result is a special case of more general approaches given in [111, 67].

- The same result can be obtained by formulating an augmented performance with a *symmetric Lagrange's multipliers matrix* $\mathbf{L}$ [111].

Theorem 5.1
Let $\mathbf{A}_{cl}(\tilde{\mathbf{u}})$ be Hurwitz and $\mathbf{P}$ and $\mathbf{L}$ be defined by Lemma 5.2. Let $\mathbf{L}_{\tilde{u}_i}$ be the solution of the following Lyapunov equation:

$$\mathbf{L}_{\tilde{u}_i}\mathbf{A}_{cl}^T(\tilde{\mathbf{u}}) + \mathbf{A}_{cl}(\tilde{\mathbf{u}})\mathbf{L}_{\tilde{u}_i} = \mathbf{L}\mathbf{c}_i^T\mathbf{b}_i^T + \mathbf{b}_i\mathbf{c}_i\mathbf{L} \tag{5.10}$$

Then the Hessian $e_{\tilde{\mathbf{u}}\tilde{\mathbf{u}}}$ is

$$e_{\tilde{\mathbf{u}}\tilde{\mathbf{u}}}(\tilde{\mathbf{u}}) = \mathbf{S} + \mathbf{S}^T \tag{5.11}$$

where $\mathbf{S} \in \mathbb{R}^{n_u \times n_u}$ is a matrix, whose i-th row is the main diagonal of

$$-2\mathbf{C}\mathbf{L}_{\tilde{u}_i}\mathbf{P}\mathbf{B} \tag{5.12}$$

Proof 5.3 It follows from the Hessian's definition that $(e_{\tilde{\mathbf{u}}\tilde{\mathbf{u}}}(\tilde{\mathbf{u}}))_{ij} = \frac{\partial}{\partial \tilde{u}_i}e_{\tilde{u}_j}(\tilde{\mathbf{u}})$. Consider $\mathbf{P}$ and $\mathbf{L}$ as matrix functions $\mathbb{R}^{n_u} \to \mathbb{R}^{n \times n}$ Lemma 5.2 states that $e_{\tilde{u}_j}(\tilde{\mathbf{u}}) = (-2\mathbf{C}\mathbf{L}\mathbf{P}\mathbf{B})_{jj} = -2\mathbf{c}_j\mathbf{L}\mathbf{P}\mathbf{b}_j$. Hence,

$$\frac{\partial}{\partial \tilde{u}_i}e_{\tilde{u}_j}(\tilde{\mathbf{u}}) = -2\frac{\partial}{\partial \tilde{u}_i}\mathbf{c}_j\mathbf{L}\mathbf{P}\mathbf{b}_j$$
$$= -2\mathbf{c}_j(\mathbf{L}_{\tilde{u}_i}\mathbf{P} + \mathbf{L}\mathbf{P}_{\tilde{u}_i})\mathbf{b}_j$$
$$= -2\mathbf{c}_j\mathbf{L}_{\tilde{u}_i}\mathbf{P}\mathbf{b}_j - 2\mathbf{c}_j\mathbf{L}\mathbf{P}_{\tilde{u}_i}\mathbf{b}_j$$

By $\mathbf{S}$'s definition, it follows that $(\mathbf{S})_{ij} \triangleq s_{ij} = -2\mathbf{c}_j\mathbf{L}_{\tilde{u}_i}\mathbf{P}\mathbf{b}_j$ and that:

$$e_{\tilde{u}_i\tilde{u}_j}(\tilde{\mathbf{u}}) = s_{ij} - 2\mathbf{c}_j\mathbf{L}\mathbf{P}_{\tilde{u}_i}\mathbf{b}_j \tag{5.13}$$

As $\mathbf{A}_{cl}(\tilde{\mathbf{u}})$ is Hurwitz, and by virtue of Eq. (5.10), $\mathbf{L}_{\tilde{u}_i}$ can be written as a Lyapunov integral. Hence s_{ij} can be written in integral form:

$$s_{ij} = -2\mathbf{c}_j\mathbf{L}_{\tilde{u}_i}\mathbf{P}\mathbf{b}_j = -2\mathbf{c}_j\left(\int_0^\infty e^{\mathbf{A}_{cl}(\tilde{\mathbf{u}})t}(\mathbf{L}\mathbf{c}_i^T\mathbf{b}_i^T + \mathbf{b}_i\mathbf{c}_i\mathbf{L})e^{\mathbf{A}_{cl}(\tilde{\mathbf{u}})^T t}dt\right)\mathbf{P}\mathbf{b}_j$$

$$= -2\int_0^\infty \Big((\mathbf{c}_j e^{\mathbf{A}_{cl}(\tilde{\mathbf{u}})t}\mathbf{L}\mathbf{c}_i^T)(\mathbf{b}_i^T e^{\mathbf{A}_{cl}(\tilde{\mathbf{u}})^T t}\mathbf{P}\mathbf{b}_j)$$
$$+ (\mathbf{c}_j e^{\mathbf{A}_{cl}(\tilde{\mathbf{u}})t}\mathbf{b}_i)(\mathbf{c}_i\mathbf{L}e^{\mathbf{A}_{cl}(\tilde{\mathbf{u}})^T t}\mathbf{P}\mathbf{b}_j)\Big)\,dt \tag{5.14}$$

In the same manner, by using Eq. (5.9), $\mathbf{P}_{\tilde{u}_i}$ can be written as a Lyapunov integral. Hence the second term in the RHS of Eq. (5.13) can also be written in an integral form:

$$-2\mathbf{c}_j\mathbf{L}\mathbf{P}_{\tilde{u}_i}\mathbf{b}_j = -2\mathbf{c}_j\mathbf{L}\left(\int_0^\infty e^{\mathbf{A}_{cl}(\tilde{\mathbf{u}})^T t}(\mathbf{P}\mathbf{b}_i\mathbf{c}_i + \mathbf{c}_i^T\mathbf{b}_i^T\mathbf{P}_i)e^{\mathbf{A}_{cl}(\tilde{\mathbf{u}})t}\mathrm{d}t\right)\mathbf{b}_j$$

$$= -2\int_0^\infty \Big((\mathbf{c}_j\mathbf{L}e^{\mathbf{A}_{cl}(\tilde{\mathbf{u}})^T t}\mathbf{P}\mathbf{b}_i)(\mathbf{c}_i e^{\mathbf{A}_{cl}(\tilde{\mathbf{u}})t}\mathbf{b}_j)$$

$$+ (\mathbf{c}_j\mathbf{L}e^{\mathbf{A}_{cl}(\tilde{\mathbf{u}})^T t}\mathbf{c}_i^T)(\mathbf{b}_i^T\mathbf{P}_i e^{\mathbf{A}_{cl}(\tilde{\mathbf{u}})t}\mathbf{b}_j)\Big)\mathrm{d}t$$

Comparing the last equation to Eq. (5.14) and recalling that scalars are invariant under transposition, concludes that $-2\mathbf{c}_j\mathbf{L}\mathbf{P}_{\tilde{u}_i}\mathbf{b}_j = s_{ji}$. Hence

$$e_{\tilde{u}_i\tilde{u}_j}(\tilde{\mathbf{u}}) = s_{ij} + s_{ji}$$

which is the ij-th element in Eq. (5.11).

Remarks.

- It follows that besides solving two Lyapunov equations for $\mathbf{L}$ and $\mathbf{P}$, computation of $e_{\tilde{\mathbf{u}}\tilde{\mathbf{u}}}(\tilde{\mathbf{u}})$ requires solving another n_u Lyapunov equations.

- A more general method for computing $e_{\mathbf{u}\mathbf{u}}(\mathbf{u})$ was provided by B. E. A. Milani [67]. However, it uses $\mathbf{P}_{\tilde{u}_i}$ rather than $\mathbf{L}_{\tilde{u}_i}$; it wasn't accompanied by any proof and it leads to a slightly different matrix. Namely, Milani's method derives the result that was obtained here multiplied by 0.5.

Lemma 5.3
The gradient and Hessian of J at a given $\tilde{\mathbf{u}}$ is:

$$J_{\tilde{\mathbf{u}}}(\tilde{\mathbf{u}}) = \frac{1}{2}e_{\tilde{\mathbf{u}}}(\tilde{\mathbf{u}}) + \tilde{\mathbf{u}}\mathbf{R}; \quad J_{\tilde{\mathbf{u}}\tilde{\mathbf{u}}}(\tilde{\mathbf{u}}) = \frac{1}{2}e_{\tilde{\mathbf{u}}\tilde{\mathbf{u}}}(\tilde{\mathbf{u}}) + \mathbf{R}$$

Proof 5.4 This is a result of a simple matrix calculus, Lemma 5.2 and Theorem 5.1.

The main computational effort, required for computing the gradient and Hessian of J at a given $\tilde{\mathbf{u}}$, sums up in solving $(n_u + 2)$ Lyapunov equations. This enables applying Newton's method for solving the optimal control problem, described in Definition 5.1. A corresponding algorithm is formulated in Fig. 5.1.

Input: $\mathbf{A}, \mathbf{B} = \begin{bmatrix} \mathbf{b}_1 & \mathbf{b}_2 & \ldots & \mathbf{b}_{n_u} \end{bmatrix}$, $\mathbf{C} = \begin{bmatrix} \mathbf{c}_1^T & \mathbf{c}_2^T & \ldots & \mathbf{c}_{n_u}^T \end{bmatrix}^T$, $\mathbf{Q} \geq 0, \mathbf{R} > 0$, $\mathbf{x}(0) \in \mathbb{R}^n$.

Initialization:

(1) Provide an initial guess $\tilde{\mathbf{u}}^0$. All the values should be positive.

(2) Select a convergence threshold - ε

Iterations: For $k = \{0, 1, 2, \ldots\}$:

(1) Solve

$$\mathbf{P}^k \mathbf{A}_{cl}^k + (\mathbf{A}_{cl}^k)^T \mathbf{P}^k = -\mathbf{Q}$$
$$\mathbf{L}^k (\mathbf{A}_{cl}^k)^T + \mathbf{A}_{cl}^k \mathbf{L}^k = -\mathbf{x}(0)\mathbf{x}(0)^T$$

for $\mathbf{P}^k$ and $\mathbf{L}^k$ where

$$\mathbf{A}_{cl}^k = \mathbf{A} - \mathbf{B} \operatorname{diag}(\tilde{\mathbf{u}}^k) \mathbf{C}$$

(2) Compute the gradient:

$$J_{\tilde{\mathbf{u}}}^k = \frac{1}{2} e_{\tilde{\mathbf{u}}}^k + \tilde{\mathbf{u}}^k \mathbf{R}$$

where $e_{\tilde{\mathbf{u}}}^k$ is a row vector, constructed from the main diagonal of $-2\mathbf{B}^T \mathbf{P}^k \mathbf{L}^k \mathbf{C}^T$.

(3) If $\|J_{\tilde{\mathbf{u}}}^k\| < \varepsilon$, stop iterating, otherwise - continue.

(4) For $i = 1, 2, \ldots, n_u$ solve:

$$\mathbf{L}_i^k (\mathbf{A}_{cl}^k)^T + \mathbf{A}_{cl}^k \mathbf{L}_i^k = \mathbf{L}^k \mathbf{c}_i^T \mathbf{b}_i^T + \mathbf{b}_i \mathbf{c}_i \mathbf{L}^k$$

and set the i-th row of $\mathbf{S}^k$ to be the main diagonal of $-2\mathbf{B}^T \mathbf{P}^k \mathbf{L}_i^k \mathbf{C}^T$.

(5) $J_{\tilde{\mathbf{u}}\tilde{\mathbf{u}}}^k = \frac{1}{2}(\mathbf{S}^k + (\mathbf{S}^k)^T) + \mathbf{R}$

(6) Compute new gains:

$$\tilde{\mathbf{u}}^{k+1} = \tilde{\mathbf{u}}^k - (J_{\tilde{\mathbf{u}}\tilde{\mathbf{u}}}^k)^{-1}(J_{\tilde{\mathbf{u}}}^k)^T$$

Output: $\tilde{\mathbf{u}}^k$.

Figure 5.1: Optimal control Problem 5.1—calculation of the optimal viscous gains.

Example 5.1.1: Free Vibration

The suggested method is used for calculating optimal VDs gains for a free vibrating 20-story model. The calculations were carried out numerically by using accordingly developed MATLAB® routines.

The 20-story model is based on that suggested by Spencer et al. [90] as a control benchmark problem for seismically excited buildings, with slight modifications. It originates from a planned north-south moment resisting frame (MRF) of a 20-story, steel structure. The structure is a typical medium-rise to high-rise building in the Los Angeles, California region. The modeled MRF is 30.48 [m] in plan, 80.77 [m] in height and it consists of 20 stories and 2 basement levels with five bays of 6.10 [m]. The floors are composite made of concrete and steel and the columns are made of 345 [MPa] steel. Typical story heights are 3.96 [m], except the two basement levels and the ground floor, which are 3.65 [m] and 5.49 [m] respectively. The columns' bases are modeled as pinned and secured to the ground. Concrete foundation walls and surrounding soil are assumed to restrain the structure at the ground level from horizontal displacement. In accordance with common structural design practice, the floor system, which provides diaphragm action, is assumed to be rigid in the horizontal plane. The floor system is composed of 248 [MPa] steel wide-flange beams acting compositely with the floor slab. The basement floor beams are simply connected to the columns. The inertial effects of each level are assumed to be carried evenly by the floor diaphragm to each perimeter MRF; hence each frame resists one half of the seismic mass associated with the entire structure. The seismic mass, associated with the MRF, is assumed to be $2.66 \cdot 10^5$ [kg] for the first level, $2.83 \cdot 10^5$ [kg] for the second, $2.76 \cdot 10^5$ [kg] for the third to 20th and $2.92 \cdot 10^5$ [kg] for the roof. The structure's natural vibration periods are {3.83, 1.33, 0.77, 0.547, 0.417, 0.333, 0.276, 0.231, 0.197, 0.172, 0.153, 0.136, 0.123, 0.112, 0.105, 0.102, 0.0941, 0.0857, 0.0788, 0.0731, 0.0636} [s]. A more detailed description of the structure and its model and the originating structure can be found in [90].

Guyan's method was used to reduce the number of DOFs to a manageable size [27]. In each floor, one horizontal DOF (at the sixth column) is defined as an active DOF. Namely, the tested model consists of 21 active DOFs. The damping matrix is determined from the reduced mass and stiffness matrices, based on a Rayleigh damping assumption. Floor inertial loads are uniformly distributed at the nodes of each respective floor assuming a consistent mass formulation. 21 VDs are assumed to be installed at each floor of the analyzed model. The dampers are numbered from 1 to 21 in

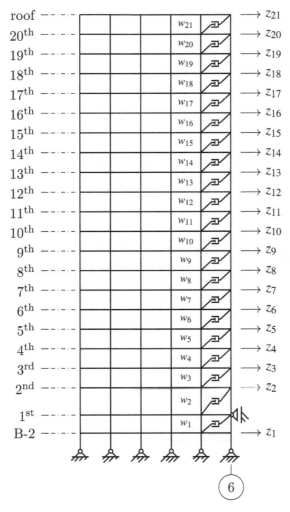

Figure 5.2: The plant's model: 20 floors evaluation model and dampers' configuration [45].

an increasing order, starting from the damper in B-2, the 1st floor and so on, until the damper in the 20th floor. The model and the dampers' configuration are shown in Fig. 5.2. In this figure z_i is the i-th DOF and w_i is the control force related to the adjacent damper. The system order is 42 and 21 viscous gains were sought. The weighting matrices were set to $\mathbf{Q} = \mathbf{I}$ and $\mathbf{R} = \mathbf{I} \cdot 10^{-14}$, each with an appropriate $\mathbf{I}$'s dimensions. The initial displacements were set to 1 and the initial velocities to zero. $\mathbf{C}$ was constructed such that $\mathbf{c}_1 \mathbf{x}(t) = \dot{z}_1(t)$, $\mathbf{c}_2 \mathbf{x}(t) = \dot{z}_2(t)$ and $\mathbf{c}_i \mathbf{x}(t) = \dot{z}_i(t) - \dot{z}_{i-1}(t)$

for all $i = 3, \ldots, 21$. The computation was initiated with initial gains equal to $\tilde{u}_{0i} = 3 \cdot 10^6$ [kg/s]. Convergence was achieved after 5 iterations. The performance index value at each iteration is described in Fig. 5.3. A comparison of the initial guess and the resulted gains is given in Fig. 5.4. Figure 5.5 describes the roof displacement of the controlled and uncontrolled models. A significant improvement is observed in the controlled plant compared to the uncontrolled one.

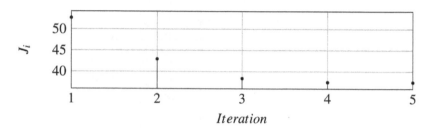

Figure 5.3: Performance index value in each iteration J_i.

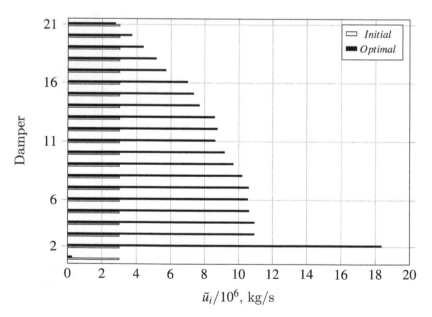

Figure 5.4: Initial and optimal gains distribution.

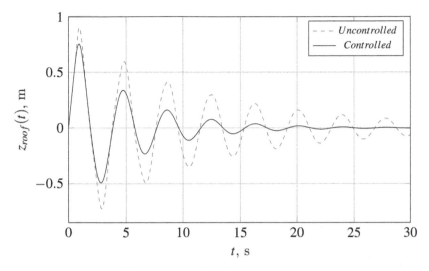

Figure 5.5: Roof displacement.

5.2 Optimal Control by Viscous Dampers: White Noise Excitation

Many VDs design methods for seismically excited structures are based on the structure's response to a given earthquake record [74], either real or artificial [76]. The disadvantage of such an approach is that it requires response simulation for the designed structure. Although alternative approaches that require no preliminary structural response simulation are also known [69], these methods are simple and suitable for structures with one dominant mode shape. Such an approach may decrease the control efficiency in complicated structures, where several modes contribute to the total response [35]. This can happen either due to the spillover effect [73] or by the simple fact that part of the structure's dominant dynamic properties is ignored. Additionally, even though these methods are usually inspired from some optimal approaches, they are not really optimal, i.e., the control design wasn't proved to satisfy some optimal condition for a given performance index. This section suggests a different design method that does not have these drawbacks.

An optimal control problem definition of an excited plant requires some additional information regarding the excitation. In many control design methods, such as LQG and active $\|H\|_2$ control, the excitation is assumed to be a stochastic white-noise signal [18]. Moreover, numerical testing shows that control designs, based on this assumption, perform well even when they are subjected to different earthquakes signals [2]. Following this approach, a modified optimal $\|H\|_2$ problem is defined below for a stable plant, excited by a white-noise ground acceleration signal and controlled by a set of linear VDs.

Problem 5.2 (H2) *An optimal control problem is defined by*

$$\dot{\mathbf{x}}(t) = \left(\mathbf{A} - \sum_{i=1}^{n_u} \mathbf{b}_i \tilde{u}_i \mathbf{c}_i\right) \mathbf{x}(t) + \mathbf{g}\ddot{z}_g(t); \quad \mathbf{x}(0) = \mathbf{0}, t \in (0, \infty) \tag{5.15a}$$

$$J(\mathbf{x}, \tilde{\mathbf{u}}) = \frac{1}{2}\bar{p}(\mathbf{x}) + \frac{1}{2}\tilde{\mathbf{u}}^T \mathbf{R}\tilde{\mathbf{u}} \tag{5.15b}$$

$$\tilde{u}_i \in \mathbb{R} \tag{5.15c}$$

where $\ddot{z}_g(t) : \mathbb{R} \to \mathbb{R}$ is a zero mean, wide-sense stationary, white noise ground acceleration signal with unitary power spectral density; $\mathbf{g} \in \mathbb{R}^n$ is a ground acceleration state input vector and

$$\bar{p}(\mathbf{x}) \triangleq E\left[\lim_{T \to \infty} \frac{1}{T} \int_0^T \mathbf{x}(t)^T \mathbf{Q}\mathbf{x}(t)\mathrm{d}t\right]$$

is an expectation of the average weighted state power. The rest of the terms are as in Problem 5.1.

Remarks.

- The assumption that the power spectral density is unitary, is not restrictive since the excitation amplitude can be always brought into account by $\mathbf{g}$.

- In contrast to Problem 5.1, the performance can't be evaluated by summing up the signals energy. Here the excitation's energy is infinite and consequently the response's energy is infinite too. Therefore, signal power is used here for evaluation, rather than energy. An expectation of $\bar{p}$ is used due to random nature of the input and the corresponding output trajectories.

The same approach that was used for solving Problem 5.1 can be adapted for solving H2.

Let $(\mathbf{x}, \tilde{\mathbf{u}})$ satisfy Eq. (5.15a). Then $\bar{p}$, which is essentially the $\|H\|_2$ norm from $\ddot{z}_g$ to $\mathbf{y} = \mathbf{Q}^{1/2}\mathbf{x}$ [111], can be computed by [28]:

$$\bar{p}(\mathbf{x}) = \mathbf{g}^T \mathbf{P}(\tilde{\mathbf{u}})\mathbf{g}$$

where $\mathbf{P}(\tilde{\mathbf{u}})$ is the solution of the above algebraic Lyapunov equation. This is verified by the following lemma.

Lemma 5.4
Let $\bar{p} : \mathscr{X} \to \mathbb{R}$ be the functional from Problem 5.2. Then

$$\bar{p}(\mathbf{x}) = \mathbf{g}^T \mathbf{P}(\tilde{\mathbf{u}}) \mathbf{g}$$

where $\mathbf{P}(\tilde{\mathbf{u}})$ is the solution of Eq. (5.8a).

This lemma is well known [18, 28], though its proof is given below for the reader's convenience.

Proof 5.5   In the following derivations, the dependency on $\tilde{\mathbf{u}}$ is omitted for clarity. Let

$$\mathbf{A}_{cl} = \mathbf{A} - \sum_{i=1}^{n_u} \mathbf{b}_i \tilde{u}_i \mathbf{c}_i$$

The state space response of the plant can be written as a convolution of the state impulse response function with the excitation:

$$\mathbf{x}(t) = e^{\mathbf{A}_{cl} t} \mathbf{g} * \ddot{z}_g(t) = \int_0^t e^{\mathbf{A}_{cl} \tau} \mathbf{g} \ddot{z}_g(t - \tau) d\tau$$

The expectation of the weighted average power is:

$$\bar{p}(\mathbf{x}) = E\left[\lim_{T \to \infty} \frac{1}{T} \int_0^T \mathbf{x}^T(t) \mathbf{Q} \mathbf{x}(t) dt \right]$$

$$= \lim_{T \to \infty} \frac{1}{T} \int_0^T E[\mathbf{x}^T(t) \mathbf{Q} \mathbf{x}(t)] dt \qquad (5.16)$$

As $\ddot{z}_g$ is a wide-sense stationary and has a zero mean:

$$E[\ddot{z}_g(t) \ddot{z}_g(t + \tau)] = \delta(\tau)$$

which turns the integrand:

$$E[\mathbf{x}^T(t) \mathbf{Q} \mathbf{x}(t)] = E\left[\left(\int_0^t \mathbf{g}^T e^{\mathbf{A}_{cl}^T \tau_1} \ddot{z}_g(t - \tau_1) d\tau_1 \right) \mathbf{Q} \left(\int_0^t e^{\mathbf{A}_{cl} \tau_2} \mathbf{g} \ddot{z}_g(t - \tau_2) d\tau_2 \right) \right]$$

$$= \mathbf{g}^T \left(\int_0^t e^{\mathbf{A}_{cl}^T \tau} \mathbf{Q} e^{\mathbf{A}_{cl} \tau} d\tau \right) \mathbf{g}$$

Substitution into Eq. (5.16) leads to:

$$\bar{p}(\mathbf{x}) = \mathbf{g}^T \left(\lim_{T \to \infty} \frac{1}{T} \int_0^T \int_0^t e^{\mathbf{A}_{cl}^T \tau} \mathbf{Q} e^{\mathbf{A}_{cl} \tau} \, \mathrm{d}\tau \mathrm{d}t \right) \mathbf{g}$$

Using Lopital's rule with respect to T, the limit reduces to

$$\bar{p}(\mathbf{x}) = \mathbf{g}^T \left(\lim_{t \to \infty} \int_0^t e^{\mathbf{A}_{cl}^T \tau} \mathbf{Q} e^{\mathbf{A}_{cl} \tau} \, \mathrm{d}\tau \right) \mathbf{g}$$

Denote the integral inside the brackets as

$$\mathbf{P}(t) \triangleq \int_0^t e^{\mathbf{A}_{cl}^T \tau} \mathbf{Q} e^{\mathbf{A}_{cl} \tau} \, \mathrm{d}\tau$$

If $\mathbf{A}_{cl}$ is Hurwitz, then $\mathbf{P}(t)$ satisfies [96]:

$$\lim_{t \to \infty} \left(\mathbf{A}_{cl}^T \mathbf{P}(t) + \mathbf{P}(t) \mathbf{A}_{cl} \right) = \lim_{t \to \infty} \int_0^t \mathrm{d} \left(e^{\mathbf{A}_{cl}^T \tau} \mathbf{Q} e^{\mathbf{A}_{cl} \tau} \right) = \lim_{t \to \infty} \left(e^{\mathbf{A}_{cl}^T t} \mathbf{Q} e^{\mathbf{A}_{cl} t} - \mathbf{Q} \right)$$
$$= -\mathbf{Q}$$

which is the well known algebraic Lyapunov equation from Eq. (5.8a).

This performance index has the same form like that described in Eq. (5.4), just $\mathbf{x}(0)$ is replaced by $\mathbf{g}$. Additionally, its minimization is carried out with relation to the same Lyapunov equation. Therefore, a solution to this problem can be found by using the same Newton's optimization method, suggested in Fig. 5.1, where the initial state is replaced by the earthquake input vector $\mathbf{g}$.

Example 5.2.1

The suggested method was applied to the same 20-story model from example 5.1.1. The vector $\mathbf{g}$ in this case is equal to 0 in the first 21 elements and -1 in the others. Four cases were analyzed:

Case 1: an uncontrolled structure.

Case 2: a structure with VDs, designed using the least-squares approximation method [69]. The method requires computation of a time history response to some reference earthquake record.

Case 3: a structure with VDs, designed by a modified viscoelastic design method [76], where the stiffness of each VE damper was set to zero such that just its viscous component contributes to the control forces. The method requires computation of time history response to an artificial white noise earthquake record.

Case 4: a structure with VDs, designed by the proposed method.

In order to achieve a reasonable comparison basis, each design was carried out such that the sum of the calculated VDs' gains was $3.22 \cdot 10^8$ [kg/s]. **R** was set to **I**. **Q** was formed such that $\mathbf{x}(t)^T \mathbf{Q} \mathbf{x}(t) = 3.3 \cdot 10^{17}(z_1(t)^2 + \sum_{i=2}^{21}(z_i(t) - z_{i-1}(t))^2))$. The initial state was set to zero. **C** was constructed as in the previous example. The gain distributions, obtained for cases 2-4, are given in Fig. 5.6. Each case was simulated for the following natural earthquake records:

- El Centro;

- Hachinohe;

- Kobe and

- Northridge.

All ground motions were scaled to a PGA of 0.3g. Figure 5.7 presents the energy density spectra of the selected earthquakes and Bode plot for the structure's roof displacement relative to the ground. As it can be seen from the figure, for all ground motions, the frequency bandwidth of the structure contains the dominant part of the spectrum. Therefore, the structure is sensitive to these earthquakes.

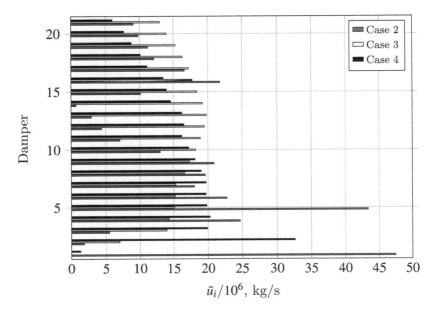

Figure 5.6: Obtained gains' distribution.

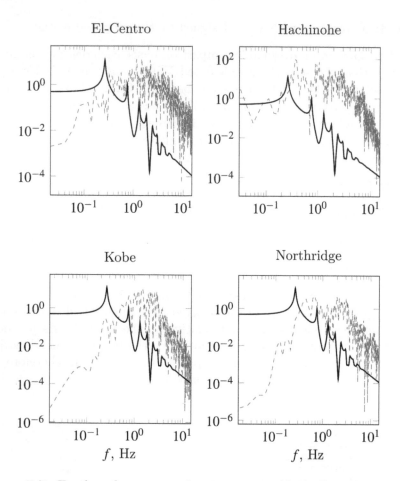

Figure 5.7: Earthquakes energy density spectra (dashed) and Bode plot for the structure's relative roof displacement (solid).

Figure 5.8 shows the roof displacements generated by each of the earthquake records. For all ground motions, a significant enhancement is evident for cases 2, 3, and 4 (30–42%) with slight advantage to case 4. The responses of cases 2, 3, and 4 are very similar, but, as it was mentioned earlier, case 4 requires no preliminary structural response analysis.

Figure 5.9 shows the peak inter-story drifts in the structure as related to the earthquake records. Similar to the effect of the control system on roof displacements, for all the ground motions, a significant improvement is evident for cases 2, 3, and 4. The structure with a control system, designed according to the proposed method (case 4), performed slightly

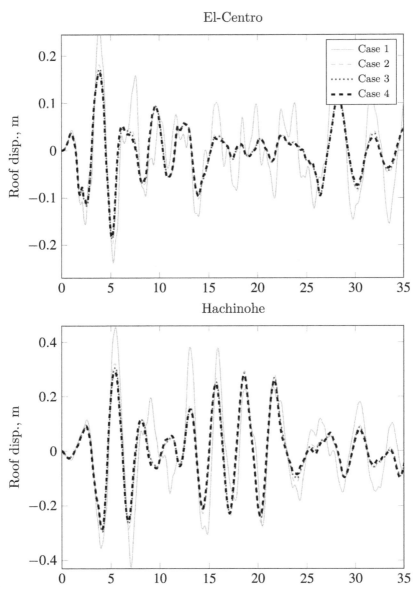

Figure 5.8: Roof displacement.

better on the lower floors; at the first floor, the improvement in case 4 was more significant, compared with cases 2 and 3. At the higher floors, the performance of case 3 is slightly better than case 4. The average reduction in peak inter-story drifts ranges between 36% and 61% for case 2, whereas for case 3, it is 40–66% and for case 4, from 44% to 68%.

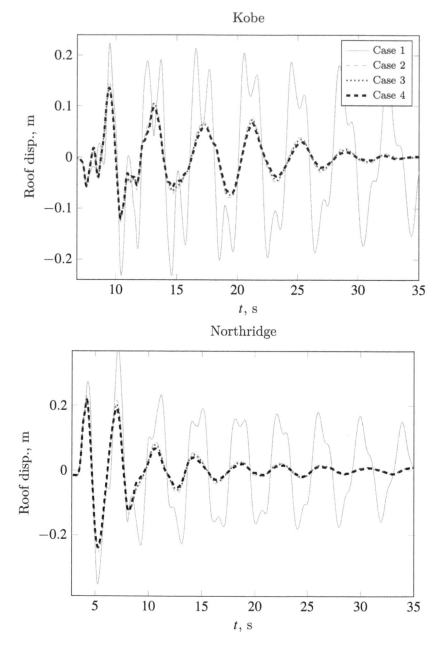

Figure 5.8(Cont.): Roof displacement.

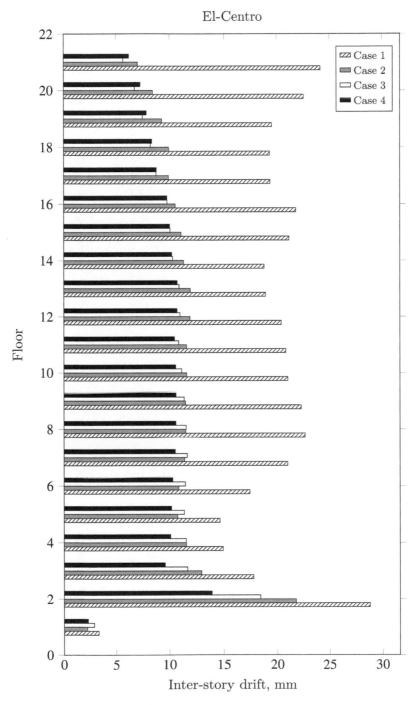

Figure 5.9: Peak inter-story drifts.

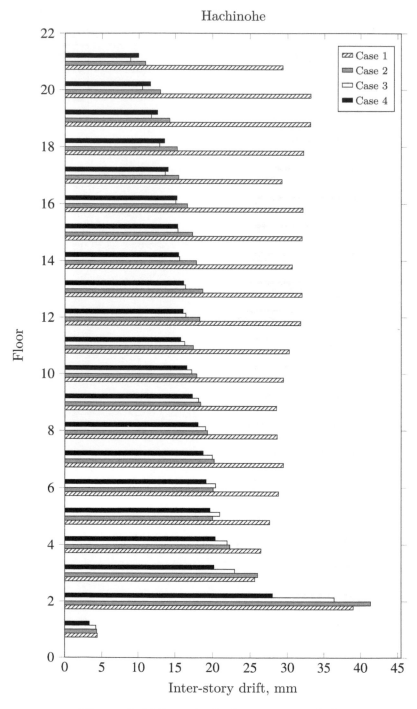

Figure 5.9(Cont.): Peak inter-story drifts.

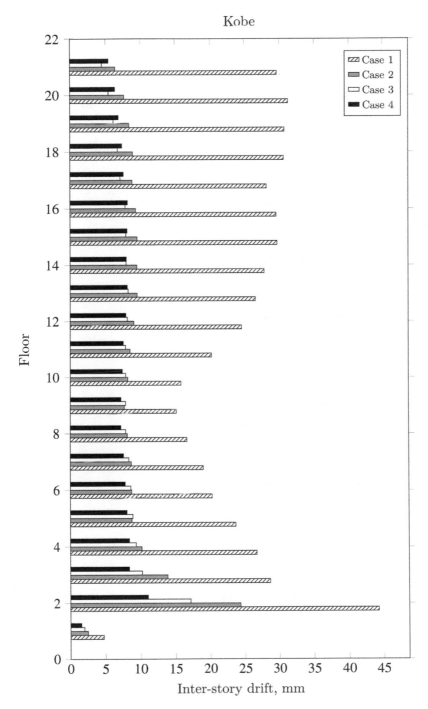

Figure 5.9(Cont.): Peak inter-story drifts.

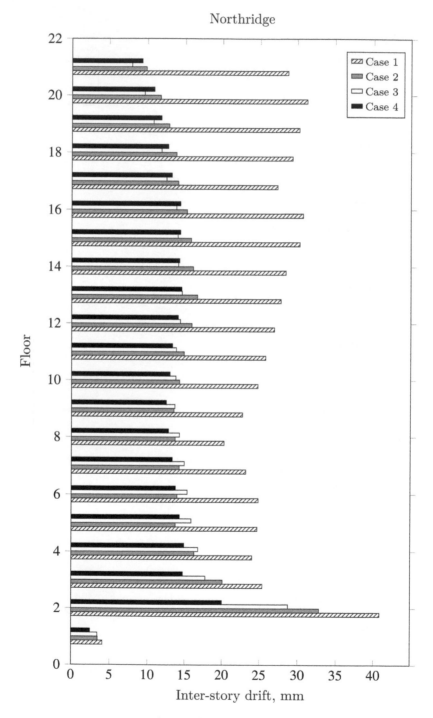

Figure 5.9(Cont.): Peak inter-story drifts.

Chapter 6

Semi-Active Control

In active control, the freedom to apply arbitrary control forces to the structure makes it a powerful tool for suppressing structural vibration. However, this approach is risky from the stability viewpoint since active control devices can potentially excite the structure. That is, even if an active control law is designed as a stabilizing one, it is not inherently stable. Practical active control implementation requires state sensing and when high order plants are addressed, it is common to use state estimation in order to reduce the number of sensors. It means that besides the basic control law, the active feedback system also involves computers, sensors, signal converters, modeling and design errors, processing time, etc., where each element contributes its own time delay and inaccuracies to the feedback system. Putting all these together with the potential embodied instability of the active actuators, there is always a concern that the closed loop system might become unstable.

Semi-active control is a well known strategy [68] that uses semi-active dampers for improving stability of dynamical systems. Unlike active control, it is incapable of adding mechanical energy to the plant, i.e., it cannot excite the plant. It also differs from passive control (in the structural control sense) as its ability to alter its dissipative properties in real time provides additional flexibility to the control law and hence extends its ability to match performance requirements.

There are studies [85], which refer to semi-active control as one that is restricted to systematic energy dissipation but allows actuators to add energy to the structure in a local sense. However, in this book, the term *semi-active control* has a more severe meaning. It refers to a control system whose actuators are restricted to local energy dissipation (see Section 2.2). That is, the semi-active actuators are passive (in the general control sense) with relation to each device's tensile force (output) and elongation rate (input). It follows that, in contrast to active actuators, semi-active control systems have no potential to destabilize the structural system [89] as they are inherently stable.

This chapter discusses optimal semi-active control design methods. Section 6.1 deals with a state space model for semi-active controlled plants, corresponding control constraints and the clipped optimal control approach. Sections 6.2 to 6.4 deal with three optimal semi-active control problems, which are addressed theoretically and demonstrated numerically.

6.1 Semi-active Control Constraints

The state equation of a free vibrating linear structure, equipped with a set of semi-active dampers, is:

$$\dot{\mathbf{x}}(t) = \mathbf{A}\mathbf{x}(t) + \mathbf{B}\mathbf{w}(t); \quad \mathbf{x}(0), \forall t \in (0, t_f)$$
$$w_i \in \mathscr{W}_i(\mathbf{x}); \quad i = 1, 2, \ldots, n_w$$

where $\mathscr{W}_i$ is a set of control force trajectories, admissible for the i-th actuator. $\mathscr{W}_i$'s $\mathbf{x}$ dependency is due to the semi-active dampers' physical nature, which sets limits on the admissible force according to the damper's state.

According to Section 2.2, each $w_i \in \mathscr{W}_i(\mathbf{x})$ must agree with three constraints:

1. w_i is always opposed to the relative velocity of the damper's anchors. In Section 2.3 this condition was referred to as a requirement for negative control power. It should be noted that some studies discuss semi-active devices, whose operation principle is based on or consists of variable stiffness [61, 80]. Apparently, as stiffness forces always resist displacements, it implies that in such devices the force direction is opposed to the relative displacement, rather than the relative velocity. However, careful inspection of the actual control force direction, which is computed by the controller, reveals that it is actually opposed to the relative velocity.

2. w_i should vanish when there is no motion in the damper. It means that, from physical considerations, no force can be generated when the relative velocity in the damper is zero.

3. For some semi-active dampers, when there is a motion in the damper there is some minimal amount of damping that it provides, even in *off-state*, i.e., when no control command is given. For example, variable viscous dampers [78] or variable orifice dampers [65].

When the relative velocity of the damper anchors can be written as a linear combination of the state elements, i.e., as $c_i x$ for some $c_i^T \in \mathbb{R}^n$, then each $w_i \in \mathcal{W}_i(x)$ should satisfy the following constraints:

C1: $w_i(t)c_i x(t) \leq 0$,

C2: $c_i x(t) = 0 \to w_i(t) = 0$,

C3: $|w_i(t)| \geq w_{i,min}(t, x(t))$

for all $t \in [0, t_f]$; and for some $w_{i,min} : \mathbb{R} \times \mathbb{R}^n \to [0, \infty)$.

Remarks.

■ The first remark for Lemma 2.1 infers that $c_i = \begin{bmatrix} 0 & \psi_i^T \end{bmatrix}$.

■ In order to prevent C2 and C3 contradiction, $w_{i,min}$ must be proper. In other words, $w_{i,min}(t, x(t)) = 0$ whenever $c_i x(t) = 0$.

■ Just to make sense, for fluid based dampers with a linear off-state, it is reasonable to set $w_{i,min}(t, x(t)) = c_{di,min}|c_i x(t)|$, whereas for friction based dampers

$$w_{i,min}(t, x(t)) = \begin{cases} 0 & , c_i x(t) = 0 \\ \beta_i & , \text{otherwise} \end{cases} ; \qquad \beta_i \geq 0 \qquad (6.1)$$

should be considered. Figures 6.1(a–c) illustrate the meaning of these constraints. Each plot describes a subset of $\mathbb{R}^2$ that consists of admissible pairs $\{(w_i(t), c_i x(t))\} \subset \mathbb{R}^2$. Figure 6.1(a) refers to a general $w_{i,min}(t, x(t))$. Figure 6.1(b) and Fig. 6.1(c) refer to $w_{i,min}(t, x(t)) = c_{di,min}|c_i x(t)|$ and Eq. (6.1), respectively.

In the control design stage, constraints C1–C3 are reflected by complex, non-linear damper models, which lead to non-linear control laws. This constrains the control trajectories shaping and boils down to a very challenging optimal control design.

An interesting approach for designing an MRD based controller is to separate it into two. The first part is a *system controller* and the second part is a *damper controller*. The system controller generates the desired force trajectory according to the dynamic response of the plant, whereas the damper controller sets the commands that are required by the damper

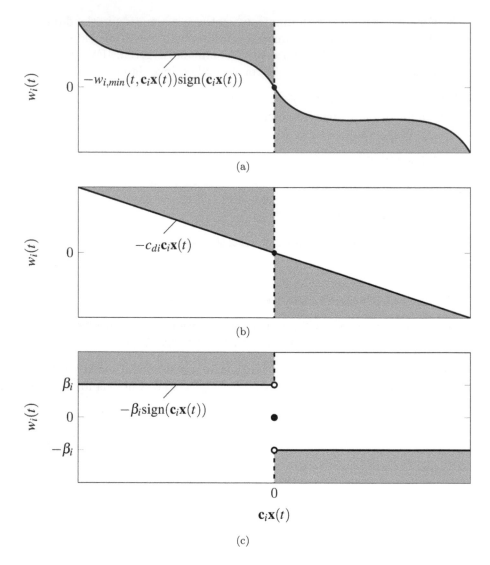

Figure 6.1: Visualizing different semi-active constraints at a given t.

for tracking this trajectory [98]. This approach is very convenient from the control design viewpoint, as it frees the designer from the need to bring into account the dependency between the damper's force and current, which is quite complex. All that is needed is to consider the bounds of the damping force. After the desired force was set, the damper controller calculates the appropriate control current by taking into account

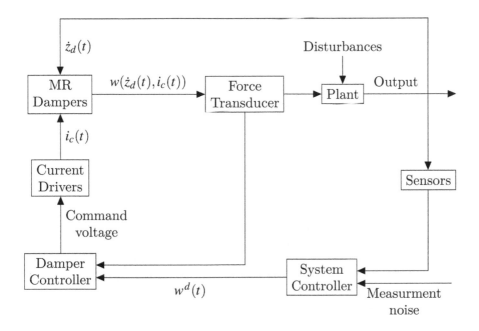

Figure 6.2: An MRD controlled system [98].

the damper's dynamics. Although that approach relates to MRD systems, it can also be used for other types of semi-active systems.

A general scheme of such control system is presented in Fig. 6.2. The branch signified by w^d is the desired control force, calculated by the system controller, and the branch signified by w is the actual force generated by the MRD devices. In several semi-active control systems, the system controller is used to compute an unbounded w^d, while the damper controller brings into account the damper's limitations and generates w that is as close as possible to w^d [5]. Clipped optimal control is an example for such approach.

Clipped Optimal Control

Clipped optimal control is a heuristic approach, in which the controller design is usually done by some known optimal method that ignores the actuator's constraints. Later, when the control forces are applied, they are arbitrarily clipped whenever the constraints are violated. This approach can be used for computing equivalent damping coefficients by three main steps. First, unbounded optimal control forces $\mathbf{w}(t)$ are calculated. Next, unbounded damping coefficients are calculated by:

$$v_i(t) = \frac{w_i(t)}{\dot{z}_d(t)}$$

where $\dot{z}_d(t)$ is the relative velocity in the semi-active damper. Finally, the bounded damping coefficient is calculated by:

$$u_i(t) = \begin{cases} u_{i,min} & ,v_i(t) \leq u_{i,min} \\ v_i(t) & ,u_{i,min} \leq v_i(t) \leq u_{i,max} \\ u_{i,max} & ,u_{i,max} \leq v_i(t) \end{cases}$$

Various forms of semi-active control that use this approach can be found in the literature. For example, in 1994, Patten et al. [71] introduced a clipped optimal control algorithm that is based on a linear quadratic regulator (LQR) with a check on the dissipation characteristics of the control force. The method's effectiveness was investigated later in another study [83]. A semi-active control methodology that uses the clipped-optimal control algorithm and a reliability-based, robust active control, was developed [108]. Pneumatic semi-active control methodology for vibration control was studied. First, the unconstrained control force is calculated by an optimal covariance control. Next, a switching logic is used to ensure that control force is applied in the appropriate direction. When the required optimal force cannot be generated in the proper direction, the controller opens a valve to minimize the generated force [80]. A clipped optimal control law was used for computing command voltage for a semi-active damper [82]. It generates rectangular pulses that cause the actual force to track the desired one. The pulses are generated in real-time by checking an ad-hoc correspondence between the desired force and the measured force. A control law for semi-active dampers, based on signal clipping of a proportional-integral controller was developed [5]. Another clipped-optimal control implementations can be found in additional works [47, 110].

The simplicity of the clipping approach is indeed a benefit, however, there is a major problem associated with it. The arbitrary clipping distorts the control trajectory and therefore raises a theoretical question on its contribution to the controlled plant.

6.2   Constrained LQR

The continuous time *constrained LQ regulator* (CLQR) problem is an optimal control problem, defined by a quadratic performance index, a linear state equation and trajectory constraints. Different formulations of CLQR problems can be found in the literature. For example, a finite horizon problem with state and control force bounds was studied [47]. An iterative algorithm was suggested for its control synthesis. A finite horizon LQR with log-barrier state constraints was reformulated as an unconstrained dynamic game [30]. The new formulation and the properties that were developed can be useful for the solution of the original CLQR problem.

This section addresses a CLQR problem with the following semi-active constraints. Each $\mathscr{W}_i(\mathbf{x})$ contains control forces that comply with constraint C1 but not necessarily C2. Additionally, $w_{min} = 0$, which turns C3 trivial. The control inputs are the control forces, i.e., $u_i = w_i$ for $i = 1,\ldots,n_w$, and the set of admissible control trajectories for the i-th actuator is denoted by $\mathscr{U}_i(\mathbf{x})$. The problem is formally defined as follows.

Definition 6.1   Let $\mathbf{x} : \mathbb{R} \to \mathbb{R}^n$ be a state trajectory and $\mathbf{u} : \mathbb{R} \to \mathbb{R}^{n_u}$ be a control trajectory. The pair $(\mathbf{x},\mathbf{u})$ is an admissible process if it satisfies the LTI state equation:

$$\dot{\mathbf{x}}(t) = \mathbf{A}\mathbf{x}(t) + \mathbf{B}\mathbf{u}(t); \quad \mathbf{x}(0), \forall t \in (0,t_f) \tag{6.2}$$

and u_i satisfies $u_i(t)\mathbf{c}_i\mathbf{x}(t) \leq 0$ for all $t \in [0,t_f]$. Here $\mathbf{A} \in \mathbb{R}^{n \times n}$; $\mathbf{B} \in \mathbb{R}^{n \times n_u}$ and $\mathbf{c}_i^T \in \mathbb{R}^n$.

Problem 6.1 (CLQR) *The constrained linear quadratic regulator (CLQR) control problem is the search after an optimal and admissible process $(\mathbf{x}^*,\mathbf{u}^*)$ that minimizes the quadratic performance index:*

$$J(\mathbf{x},\mathbf{u}) = \frac{1}{2} \int\limits_0^{t_f} \left(\mathbf{x}(t)^T \mathbf{Q}\mathbf{x}(t) + \sum_{i=1}^{n_u} u_i(t)^2 r_i \right) dt \tag{6.3}$$

where $0 \leq \mathbf{Q} \in \mathbb{R}^{n \times n}$ and $r_i > 0$ for $i = 1,\ldots,n_u$.

Before addressing the CLQR problem, some practical issues, related to the CLQR definition, should be discussed. As the defined CLQR problem ignores constraint C2, its solution might yield control forces that are not fully realizable by semi-active dampers. Namely, the controller might command the damper to generate a force when there is no relative velocity in the device. This is impossible and therefore the relevance of the defined constraint to the semi-active control problem is questionable, at least from practical viewpoint. Fortunately, this issue does not pose a significant problem. It will be shown below that by setting $u_i(t) = 0$ whenever $\mathbf{c}_i\mathbf{x}(t) = 0$, the process cost is preserved, on the one hand, but the control trajectory becomes realizable, on the other hand. Therefore, when **u** is the optimal one, such arbitrary clipping still give us a proper CLQR optimum.

This fact is established by the following two lemmas. They show that if $(\mathbf{A}, \mathbf{c}_i)$ are observable, a situation in which $\mathbf{c}_i\mathbf{x}(t) = 0$ happens to be exactly one of the following:

- A momentary situation.

- A persistent situation. In other words, it implies that the response has ended.

Let $N_\varepsilon(t)$ be a *neighborhood of t*. That is, a set $\{\tau\}$ whose members satisfy $|\tau - t| < \varepsilon$, for a given ε.

Lemma 6.1 [39]
Let the state equation of a controlled plant be

$$\dot{\mathbf{x}}(t) = (\mathbf{A} - \mathbf{BG}(t))\mathbf{x}(t); \quad \mathbf{x}(0), \forall t \in (0, t_f)$$

where $\mathbf{x} : [0, t_f] \to \mathbb{R}^n$ is a state trajectory and $\mathbf{G} : [0, t_f] \to \mathbb{R}^{n_u \times n}$ is a time varying feedback matrix.

Let $\mathbf{c}^T \in \mathbb{R}^n$. If the pair $(\mathbf{A}, \mathbf{c})$ is observable and $\mathbf{cx}(t') = 0$ for some time instance t'. Then the following two statements are equivalent:

(a) There exist some $\varepsilon > 0$ and $N_\varepsilon(t')$ such that $\mathbf{cx}(\tau) = 0$ for all $\tau \in N_\varepsilon(t')$.

(b) $\mathbf{x}(t') = \mathbf{0}$.

Proof 6.1   Let $\mathbf{A}_{cl}(t) = \mathbf{A} - \mathbf{BG}(t)$ and let $\mathbf{n}_i(t) \triangleq \mathbf{n}_{i-1}(t)\mathbf{A}_{cl}(t) + \dot{\mathbf{n}}_{i-1}(t)$ where $\mathbf{n}_0(t) = \mathbf{c}$. It follows that

$$\frac{\mathrm{d}}{\mathrm{d}t}(\mathbf{n}_i(t)\mathbf{x}(t)) = \mathbf{n}_i(t)\mathbf{A}_{cl}(t)\mathbf{x}(t) + \dot{\mathbf{n}}_i(t)\mathbf{x}(t)$$

$$= \mathbf{n}_{i+1}(t)\mathbf{x}(t)$$

Assume that (a) holds. In that case

$$\left(\frac{\mathrm{d}^{(i)}}{\mathrm{d}t^{(i)}}(\mathbf{cx}(t)) \right)\Big|_{t'} = 0$$

for all $i = 1, 2, \dots$. This yields the set of homogeneous equations:

$$\mathbf{cx}(t') \triangleq \mathbf{n}_0(t')\mathbf{x}(t') = 0$$

$$\left(\frac{d}{dt}(\mathbf{cx}(t)) \right)\Big|_{t'} = \left(\frac{d}{dt}(\mathbf{n}_0(t)\mathbf{x}(t)) \right)\Big|_{t'} = \mathbf{n}_1(t')\mathbf{x}(t') = 0$$

$$\left(\frac{d^{(2)}}{dt^{(2)}}(\mathbf{cx}(t)) \right)\Big|_{t'} = \left(\frac{d}{dt}(\mathbf{n}_1(t)\mathbf{x}(t)) \right)\Big|_{t'} = \mathbf{n}_2(t')\mathbf{x}(t') = 0$$

$$\left(\frac{d^{(3)}}{dt^{(3)}}(\mathbf{cx}(t)) \right)\Big|_{t'} = \left(\frac{d}{dt}(\mathbf{n}_2(t)\mathbf{x}(t)) \right)\Big|_{t'} = \mathbf{n}_3(t')\mathbf{x}(t') = 0$$

$$\vdots$$

In matrix form, these equations become

$$\begin{bmatrix} \mathbf{n}_0(t')\mathbf{x}(t') \\ \mathbf{n}_1(t')\mathbf{x}(t') \\ \mathbf{n}_2(t')\mathbf{x}(t') \\ \mathbf{n}_3(t')\mathbf{x}(t') \\ \vdots \end{bmatrix} = \begin{bmatrix} \mathbf{n}_0(t') \\ \mathbf{n}_1(t') \\ \mathbf{n}_2(t') \\ \mathbf{n}_3(t') \\ \vdots \end{bmatrix} \mathbf{x}(t') = \mathbf{O}(t')\mathbf{x}(t') = \mathbf{0}$$

In the linear-time-varying systems' theory, $\mathbf{O}$ is known as the observability matrix of $(\mathbf{A}_{cl}(t), \mathbf{c})$ [86]. Its rank is full for all t as long as $(\mathbf{A}_{cl}(t), \mathbf{n}_0(t))$ is observable for all t. Though, since: (1) $(\mathbf{A}, \mathbf{c})$ is observable, (2) $\mathbf{n}_0(t') = \mathbf{c}$ and (3) it is known that a feedback does not change the observability of a closed loop system [97], then $(\mathbf{A}_{cl}(t), \mathbf{c})$ is observable too and $\mathbf{O}(t)$'s rank is n for all t. Hence, these equations have a unique solution - $\mathbf{x}(t') = \mathbf{0}$. The converse is obvious.

Remarks.

- The given plant has no external inputs. Hence, $\mathbf{x}(t') = \mathbf{0}$ infers that $\mathbf{x}(\tau) = \mathbf{0}$ for all $\tau \geq t'$, by virtue of the system's state definition. This means that the response has ended and no control effort is needed any more.

- Let a punctured neighborhood be $\bar{N}_\varepsilon(t') = N_\varepsilon(t') \setminus t'$. When $\mathbf{cx}(t') = 0$ but $\mathbf{x}(t') \neq 0$, the lemma assures that there exist $\varepsilon > 0$ such that $\mathbf{cx}(\tau) \neq 0$ for any $\tau \in \bar{N}_\varepsilon(t')$.

- Assume that the response ends after t_f. It follows that, $\mathbf{x}(t) \neq \mathbf{0}$ for all $t \in [0, t_f]$. Let $T = \{t_i | t_i \in [0, t_f], \mathbf{cx}(t_i) = 0\}$ be a set of time instances in which $\mathbf{cx}$ vanishes. It follows that T is at most countable.

Lemma 6.2

Let $(\mathbf{x},\mathbf{u})$ and $(\mathbf{x}_0,\mathbf{u}_0)$ be two processes that satisfy Eq. (6.2) and let J be defined by Eq. (6.3). If $T = \{t_i|\mathbf{u}(t_i) \neq \mathbf{u}_0(t_i), t_i \in [0,t_f]\}$ is at most countable, then $\mathbf{x}_0 = \mathbf{x}$ and $J(\mathbf{x},\mathbf{u}) = J(\mathbf{x}_0,\mathbf{u}_0)$.

Proof 6.2 $\mathbf{x}$ and $\mathbf{x}_0$ satisfy

$$\mathbf{x}(t) = \mathbf{x}(0) + \int_0^t e^{\mathbf{A}(t-\tau)}\mathbf{B}\mathbf{u}(\tau)d\tau; \quad \mathbf{x}_0(t) = \mathbf{x}(0) + \int_0^t e^{\mathbf{A}(t-\tau)}\mathbf{B}\mathbf{u}_0(\tau)d\tau$$

for every $t \in [0,t_f]$. Here $e^{\mathbf{A}t}$ is the state transition matrix. Hence

$$\mathbf{x}(t) - \mathbf{x}_0(t) = \int_0^t e^{\mathbf{A}(t-\tau)}\mathbf{B}(\mathbf{u}(\tau) - \mathbf{u}_0(\tau))d\tau$$

As $\mathbf{u}(t) - \mathbf{u}_0(t) = \mathbf{0}$ for any $t \notin T$, we get

$$\mathbf{x}(t) - \mathbf{x}_0(t) = \sum_{t_i \in T}\left(\int_{t_i}^{t_i} e^{\mathbf{A}(t-\tau)}\mathbf{B}(\mathbf{u}(\tau) - \mathbf{u}_0(\tau))d\tau\right) = \mathbf{0}$$

Similarly:

$$J(\mathbf{x},\mathbf{u}) - J(\mathbf{x}_0,\mathbf{u}_0) = \frac{1}{2}\int_0^{t_f}\left(\mathbf{x}(t)^T\mathbf{Q}\mathbf{x}(t) + \sum_{i=1}^{n_u}u_i^2(t)r_i\right)dt$$

$$-\frac{1}{2}\int_0^{t_f}\left(\mathbf{x}_0(t)^T\mathbf{Q}\mathbf{x}_0(t) + \sum_{i=1}^{n_u}u_{0i}^2(t)r_i\right)dt$$

$$=\frac{1}{2}\int_0^{t_f}(\mathbf{x}(t) + \mathbf{x}_0(t))^T\mathbf{Q}(\mathbf{x}(t) - \mathbf{x}_0(t))dt$$

$$+\frac{1}{2}\sum_{t_i \in T}\left(\int_{t_i}^{t_i}\sum_{i=1}^{n_u}(u_i^2(t) - u_{0i}^2(t))r_idt\right)$$

$$=0$$

Let $(\mathbf{x}^*,\mathbf{u}^*)$ be a CLQR optimum and let $(\mathbf{x}_c^*,\mathbf{u}_c^*)$ be a clipped process such that

$$u_{ci}^*(t) = \begin{cases} 0 & \mathbf{c}_i\mathbf{x}(t) = 0 \\ u_i^*(t) & \text{otherwise} \end{cases}$$

where $(\mathbf{A}, \mathbf{c}_i)$ is observable for $i = 1, \ldots, n_u$. Assume that the response ends after t_f. Then, by Lemma 6.1, $T_i = \{t | \mathbf{c}_i \mathbf{x}(t) = 0, t \in [0, t_f]\}$ is at most countable. It means that $\mathbf{u}_c^*$ differs from $\mathbf{u}^*$ at the set of time instances - $\cup_{i=1}^{n_u} T_i$, which is at most countable. By Lemma 6.2, $J(\mathbf{x}^*, \mathbf{u}^*) = J(\mathbf{x}_c^*, \mathbf{u}_c^*)$, which concludes that the clipped signal is still a proper optimum. Physically, it means that the total contribution of the control force at these time instances is insignificant and can be ignored.

Hence, the fact that $u_i(t) \neq 0$ when $\mathbf{c}_i \mathbf{x}(t) = 0$, does not pose a practical issue and the CLQR problem can be used to define an optimal control problem of semi-active controllers.

The following lemma and theorem provide necessary conditions for a CLQR optimum.

Lemma 6.3
If $(\mathbf{x}^*, \mathbf{u}^*)$ is a CLQR optimum then the following equations hold:

$$\dot{\mathbf{x}}^*(t) = \mathbf{A}\mathbf{x}^*(t) + \mathbf{B}\mathbf{u}^*(t); \quad \mathbf{x}^*(0), \forall t \in (0, t_f) \tag{6.4a}$$

$$\dot{\mathbf{p}}^*(t) = -\mathbf{A}^T \mathbf{p}^*(t) - \mathbf{Q}\mathbf{x}^*(t) - \sum_{i=1}^{n_u} \mu_i^*(t) \mathbf{c}_i^T u_i^*(t); \quad \mathbf{p}^*(t_f) = \mathbf{0}, \forall t \in (0, t_f) \tag{6.4b}$$

$$0 = r_i u_i^*(t) + \mathbf{b}_i^T \mathbf{p}^*(t) + \mu_i^*(t) \mathbf{c}_i \mathbf{x}^*(t); \quad i = 1, \ldots, n_u \tag{6.4c}$$

$$\mu_i^*(t) \begin{cases} \geq 0 & , \mathbf{c}_i \mathbf{x}^*(t) u_i^*(t) = 0 \\ = 0 & , \mathbf{c}_i \mathbf{x}^*(t) u_i^*(t) < 0 \end{cases} \tag{6.4d}$$

Here $\mathbf{p}^* : \mathbb{R} \to \mathbb{R}^n$ is the costate trajectory; $\mathbf{b}_i$ is the i-th column of $\mathbf{B}$ and $\mu_i : \mathbb{R} \to [0, \infty)$ is the i-th KKT multiplier.

Proof 6.3   Consider the notations used in Section 3.3 in the definition of KKT's conditions. The set of inequality constraints, which is related to the CLQR problem, is $f_i^c(t, \mathbf{x}(t), \mathbf{u}(t)) = u_i(t) \mathbf{c}_i \mathbf{x}(t)$ for $i = 1, \ldots, n_u$; $\mathbf{f}(t, \mathbf{x}(t), \mathbf{u}(t)) = \mathbf{A}\mathbf{x}(t) + \mathbf{B}\mathbf{u}(t)$ is the state space mapping; and $l(t, \mathbf{x}(t), \mathbf{u}(t)) = 0.5\mathbf{x}(t)^T \mathbf{Q}\mathbf{x}(t) + 0.5\sum_{i=1}^{n_u} u_i(t)^2 r_i$. Let the hypothesis hold. It follows from KKT's conditions that $(\mathbf{x}^*, \mathbf{u}^*)$ must satisfy Eqs. (3.15). Differentiating the i-th inequality constraint with respect to $\mathbf{x}$ and $\mathbf{u}$ yields

$$\frac{\partial}{\partial \mathbf{x}} f_i^c(t, \mathbf{x}(t), \mathbf{u}(t)) = u_i(t) \mathbf{c}_i$$

$$\frac{\partial}{\partial \mathbf{u}} f_i^c(t, \mathbf{x}(t), \mathbf{u}(t)) = \begin{bmatrix} \ldots & 0 & \mathbf{c}_i \mathbf{x}(t) & 0 & \ldots \end{bmatrix} \in \mathbb{R}^{1 \times n_u}$$
$$\uparrow$$
$$\text{the } i\text{-th}$$
$$\text{element}$$

respectively. Additionally:

$$\mathbf{f_x}(t, \mathbf{x}(t), \mathbf{u}(t)) = \mathbf{A}^T; \quad \mathbf{f_u}(t, \mathbf{x}(t), \mathbf{u}(t)) = \mathbf{B}^T$$

$$l_\mathbf{x}(t, \mathbf{x}(t), \mathbf{u}(t)) = \mathbf{x}(t)^T \mathbf{Q}; \quad l_\mathbf{u}(t, \mathbf{x}(t), \mathbf{u}(t)) = (r_i u_i(t))_{i=1}^{n_u}$$

All that is left is to substitute into Eq. (3.15).

Theorem 6.1
If $(\mathbf{x}^*, \mathbf{u}^*)$ *is a CLQR optimum, then it satisfies:*

$$\dot{\mathbf{x}}^*(t) = \mathbf{A}\mathbf{x}^*(t) + \mathbf{B}\mathbf{u}^*(t); \quad \mathbf{x}^*(0), \forall t \in (0, t_f)$$

$$u_i^*(t) = -\frac{\mathbf{b}_i^T \mathbf{P}(t)\mathbf{x}^*(t)}{r_i} \alpha_i(t)$$

where $\mathbf{P} : \mathbb{R} \to \mathbb{R}^{n \times n}$ *and* $\alpha_i : \mathbb{R} \to \{0, 1\}$ *are defined by*

$$\dot{\mathbf{P}}(t) = -\mathbf{A}^T \mathbf{P}(t) - \mathbf{P}(t)\mathbf{A} - \mathbf{Q} + \mathbf{P}(t)\mathbf{B}\operatorname{diag}((\alpha_i(t)/r_i)_{i=1}^{n_u})\mathbf{B}^T \mathbf{P}(t)$$

$$\mathbf{P}(t_f) = \mathbf{0}; \quad \forall t \in (0, t_f) \tag{6.5}$$

$$\alpha_i(t) = \begin{cases} 1 & , \operatorname{sign}(\mathbf{c}_i \mathbf{x}^*(t)) = \operatorname{sign}(\mathbf{b}_i^T \mathbf{P}(t)\mathbf{x}^*(t)) \ or \ \mathbf{c}_i \mathbf{x}^*(t) = 0 \\ 0 & , \operatorname{sign}(\mathbf{c}_i \mathbf{x}^*(t)) \neq \operatorname{sign}(\mathbf{b}_i^T \mathbf{P}(t)\mathbf{x}^*(t)) \end{cases} \tag{6.6}$$

Proof 6.4   If $(\mathbf{x}^*, \mathbf{u}^*)$ is an optimum, then by Lemma 6.3, it satisfies Eqs. (6.4) for any $t \in [0, t_f]$. It follows from Eqs. (6.4c) and (6.4d) that each $u_i^*(t)$ and $\mu_i^*(t)$ must fall into one of the next three cases:

1. $u_i^*(t)\mathbf{c}_i\mathbf{x}^*(t) < 0$. Hence $\mu_i^*(t) = 0$ and:

$$0 = r_i u_i^*(t) + \mathbf{b}_i^T \mathbf{p}^*(t) \ \to \ u_i^*(t) = -\frac{\mathbf{b}_i^T \mathbf{p}^*}{r_i}$$

Additionally, $u_i(t)\mathbf{c}_i\mathbf{x}^*(t) < 0$ implies that:

$$\mathbf{c}_i\mathbf{x}^*(t)\left(-\frac{\mathbf{b}_i^T \mathbf{p}^*}{r_i}\right) < 0 \ \to \ \mathbf{c}_i\mathbf{x}^*(t)\mathbf{b}_i^T \mathbf{p}^*(t) > 0$$

$$\Rightarrow \operatorname{sign}(\mathbf{c}_i\mathbf{x}^*(t)) = \operatorname{sign}(\mathbf{b}_i^T \mathbf{p}^*(t))$$

Note that in this case $u_i^*(t)$ coincides with the unconstrained optimum.

2. $u_i^*(t)\mathbf{c}_i\mathbf{x}^*(t) = 0$ and $\mathbf{c}_i\mathbf{x}^*(t) \neq 0$, i.e., the optimum is on the boundary of the constraining subset. It follows that $\mu_i^*(t) \geq 0$, $u_i^*(t) = 0$ and:

$$0 = \mathbf{b}_i^T \mathbf{p}^*(t) + \mu_i^*(t)\mathbf{c}_i\mathbf{x}^*(t)$$

$$\to \mu_i^*(t) = -\frac{\mathbf{b}_i^T \mathbf{p}^*}{\mathbf{c}_i\mathbf{x}^*(t)}$$

and $\mu_i^*(t) \geq 0$ implies that:

$$\operatorname{sign}(\mathbf{c}_i\mathbf{x}^*(t)) \neq \operatorname{sign}(\mathbf{b}_i^T \mathbf{p}^*(t))$$

3. $u_i^*(t)\mathbf{c}_i\mathbf{x}^*(t) = 0$ and $\mathbf{c}_i\mathbf{x}^*(t) = 0$, i.e., the optimum is on the boundary of the constraining subset and $\mu_i^*(t) \geq 0$, again. However, now it follows that

$$0 = r_i u_i^*(t) + \mathbf{b}_i^T \mathbf{p}^*(t) \rightarrow u_i^*(t) = -\frac{\mathbf{b}_i^T \mathbf{p}^*(t)}{r_i}$$

which is an unconstrained optimum. According KKT's theorem, when the unconstrained optimum is an admissible one, the inequality constraint should be omitted. Since $f_i^c(\mathbf{x}^*(t), u_i^*(t)) \leq 0$ is satisfied and $u_i^*(t)$ is the unconstrained optimum (as in case 1), it can be concluded that $\mu_i^*(t) = 0$.

Hence, the control force can now be written as

$$u_i^*(t) = -\frac{\mathbf{b}_i^T \mathbf{p}^*(t)}{r_i} \alpha_i(t) \tag{6.7}$$

where α_i is given by

$$\alpha_i(t) = \begin{cases} 1 & , \text{sign}(\mathbf{c}_i\mathbf{x}^*(t)) = \text{sign}(\mathbf{b}_i^T \mathbf{p}^*(t)) \text{ or } \mathbf{c}_i\mathbf{x}^*(t) = 0 \\ 0 & , \text{otherwise} \end{cases} \tag{6.8}$$

It turns out that at each time instance, either $u_i^*(t) = 0$ or $\mu_i^*(t) = 0$. Hence $u_i^*(t)\mu_i^*(t) = 0$ for any $t \in [0, t_f]$ and Eq. (6.4b) becomes:

$$\dot{\mathbf{p}}^*(t) = -\mathbf{A}^T \mathbf{p}^*(t) - \mathbf{Q}\mathbf{x}^*(t); \quad \mathbf{p}^*(t_f) = \mathbf{0}, \forall t \in (0, t_f)$$

Substituting Eq. (6.7) and then letting $\mathbf{p}^*(t) = \mathbf{P}(t)\mathbf{x}^*(t)$ leads to Eq. (6.5). The terminal condition $\mathbf{p}^*(t_f) = \mathbf{P}(t_f)\mathbf{x}^*(t_f) = \mathbf{0}$ implies $\mathbf{P}(t_f) = \mathbf{0}$. Equation 6.6 is obtained by substituting $\mathbf{p}^*(t) = \mathbf{P}(t)\mathbf{x}^*(t)$ into Eq. (6.8).

Remarks.

- The necessary condition, given in Theorem 6.1, defines a two-point-boundary-value problem whose solution is a candidate optimum for the CLQR problem.

- The KKT multipliers, i.e., $\mu_i(t)$, vanish from the equations. However, their values are

$$\mu_i^*(t) = \begin{cases} (\alpha_i(t) - 1)[\mathbf{b}_i^T \mathbf{P}\mathbf{x}^*(t)]/(\mathbf{c}_i\mathbf{x}^*(t)) & \mathbf{c}_i\mathbf{x}^*(t) \neq 0 \\ 0 & \mathbf{c}_i\mathbf{x}^*(t) = 0 \end{cases}$$

- An interesting observation is that when $\alpha_i(t) = 0$ for all $i = 1, \ldots, n_u$ Eq. (6.5) is a differential Lyapunov equation and when $\alpha_i(t) = 1$ for all $i = 1, \ldots, n_u$ it is a differential Riccati's equation.

Example 6.2.1—Multi input CLQR

A CLQR design was carried out for a dynamic model of a 3 floors, free vibrating, reinforced concrete (RC), plane frame. The concrete elastic modulus was assumed to be 30 [GPa]. The columns and beams sections are 40×30 [cm] and 60×30 [cm], respectively. It is assumed that the mass is concentrated at each level's floor and it brings into account dead and live loads. For the 1st and 2nd floors the mass is 23.2 [ton] and for the 3rd floor it is 11.6 [ton]. All the links are rigid. The control forces were applied by two semi-active dampers, installed in the first and second floors. All computations were carried out numerically by accordingly written MATLAB® routines. The model and its dynamic scheme are given in Fig. 6.3.

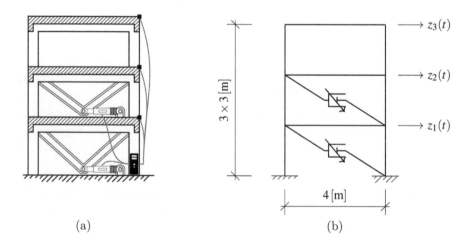

(a)   (b)

Figure 6.3: The plant's model: (a) Model illustration. (b) Dynamic scheme of the evaluated model.

The mass, damping, stiffness and control force distribution matrices are [43]:

$$\mathbf{M} = \begin{bmatrix} 22.6 & 0.345 & -0.0443 \\ 0.345 & 22.6 & 0.208 \\ -0.0443 & 0.208 & 8.62 \end{bmatrix} \cdot 10^3; \quad \mathbf{K} = \begin{bmatrix} 79.6 & -41.7 & 4.03 \\ -41.7 & 74.9 & -36.7 \\ 4.03 & -36.7 & 32.5 \end{bmatrix} \cdot 10^6$$

$$\mathbf{C}_d = \begin{bmatrix} 136 & -53.1 & 5.12 \\ -53.1 & 130 & -47 \\ 5.12 & -47 & 54.6 \end{bmatrix} \cdot 10^3; \quad \mathbf{\Phi} = \begin{bmatrix} 0.971 & -0.985 \\ 0.015 & 0.973 \\ -0.00135 & 0.0133 \end{bmatrix}$$

Finite element method and Ritz model reductions [90] were used for computing $\mathbf{K}$, $\mathbf{M}$ and $\mathbf{\Psi}$. The model natural frequencies are $\{2.85, 8.36, 12.5\}$ [Hz]. $\mathbf{C}_d$ is a Rayleigh proportional damping matrix, computed assuming that $\xi_1 = \xi_2 = 5\%$. The row vector that projects the state on the relative velocity of the i-*th* damper is $\mathbf{c}_i = \begin{bmatrix} \mathbf{0} & \boldsymbol{\psi}_i^T \end{bmatrix}^T$ where $\boldsymbol{\psi}_i$ is the i-th column of $\mathbf{\Psi}$.

The state and control weightings are:

$$\mathbf{Q} = \begin{bmatrix} 6 & -3 & 0 & 0 & 0 & 0 \\ -3 & 6 & -3 & 0 & 0 & 0 \\ 0 & -3 & 3 & 0 & 0 & 0 \\ 0 & 0 & 0 & 0 & 0 & 0 \\ 0 & 0 & 0 & 0 & 0 & 0 \\ 0 & 0 & 0 & 0 & 0 & 0 \end{bmatrix} \cdot 10^{15}; \; r = 1 \qquad (6.9)$$

The initial state vector is $\mathbf{x}(0) = \begin{bmatrix} 0 & 0 & 0 & 0.4 & 0.4 & 0.4 \end{bmatrix}^T$.

The problem was solved by a steepest descent algorithm, described in Fig. 6.4. The algorithm computes $\mathbf{P}$ for a discretized time axis by the steepest descent method. Figure 6.5 shows P_{11}'s and α_1's values. The blue regions in the graph represent time instances when $\alpha_1(t) = 1$. The white regions are time instances when $\alpha_1(t) = 0$. The two red lines indicate values of a corresponding element, obtained by solving algebraic Riccati (ARE) and Lyapunov equations. The values of $f_i^c(\mathbf{x}, \mathbf{u}) = u_i \mathbf{c}_i \mathbf{x}$ for the obtained solution are given in Fig. 6.6. It can be seen that $f_i^c(t, \mathbf{x}(t), \mathbf{u}(t)) \leq 0$ during the control process, i.e., the constraints are satisfied. Figure 6.7 shows the constrained control trajectories. Zero values are observed when the corresponding α_i is zero.

Input:
A, B, C and $\mathbf{x}(0)$.
Initialization:

(1) Construct a discrete time axis - $(t_k)_{k=1}^N$ where $t_N = t_f$.

(2) Set $\mathbf{u}_0(t_k) = 0$ for $k = 1, \ldots, N$.

(3) Define some norm $\|\cdot\| : \mathbb{R}^N \to \mathbb{R}$.

(4) Select a convergence threshold $\varepsilon > 0$ and a steepest descent coefficient $\tau > 0$.

Iterations: For $i = 0, 1, \ldots$:

(1) Solve:

$$\dot{\mathbf{x}}_i(t_k) = \mathbf{A}\mathbf{x}_i(t_k) + \mathbf{B}\mathbf{u}_i(t_k); \quad \mathbf{x}(0), k = 1, \ldots, N$$

(2) Solve:

$$\dot{\mathbf{p}}_i(t_k) = -\mathbf{A}^T\mathbf{p}_i(t_k) - \mathbf{Q}\mathbf{x}_i(t_k); \quad \mathbf{p}_i(t_N) = \mathbf{0}, k = 1, \ldots, N$$

(3) For $k = 1, \ldots, N$ and $j = 1, \ldots, n_u$, compute:

$$\alpha_{ij}(t_k) = \begin{cases} 1 & \text{sign}(\mathbf{c}_i^T\mathbf{x}(t_k)) = \text{sign}(\mathbf{b}_i^T\mathbf{p}(t_k)) \text{ or } \mathbf{c}_i^T\mathbf{x}(t_k) = 0 \\ 0 & \text{sign}(\mathbf{c}_i^T\mathbf{x}(t_k)) \neq \text{sign}(\mathbf{b}_i^T\mathbf{p}(t_k)) \end{cases}$$

and

$$H_{\mathbf{u}}(t_k, \mathbf{x}(t_k), \mathbf{p}_i(t_k), \mathbf{u}_i(t_k)) = \mathbf{u}_i(t_k)^T \, \text{diag}((r_j)_{j=1}^{n_u}) + \text{diag}((\alpha_{ij}(t_k))_{j=1}^{n_u})\mathbf{B}^T\mathbf{p}_i(t_k)$$

(4) If $\|H_{\mathbf{u}}\| > \varepsilon$ then update:

$$\mathbf{u}_{i+1}(t_k) = \mathbf{u}_i(t_k) - \tau H_{\mathbf{u}}(t_k, \mathbf{x}(t_k), \mathbf{p}_i(t_k), \mathbf{u}_i(t_k))^T$$

for $k = 1, \ldots, N$. Otherwise, set $(\alpha_j)_{j=1}^{n_u} = (\alpha_{ij})_{j=1}^{n_u}$ and stop iterating.

Solve:

$$\dot{\mathbf{P}}(t_k) = -\mathbf{A}^T\mathbf{P}(t_k) - \mathbf{P}(t_k)\mathbf{A} - \mathbf{Q}$$
$$+ \mathbf{P}(t_k)\mathbf{B}\,\text{diag}((\alpha_j(t_k)/r_j)_{j=1}^{n_u})\mathbf{B}^T\mathbf{P}(t_k); \quad \mathbf{P}(t_N) = \mathbf{0}$$

for $k = 1, \ldots, N$.
Output: P.

Figure 6.4: CLQR—a steepest descent algorithm for computing **P**.

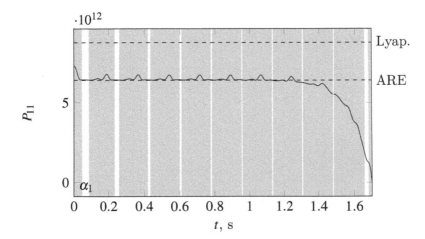

Figure 6.5: Values of P_{11}, α_1 and a corresponding element obtained by solving algebraic Riccati and Lyapunov equations.

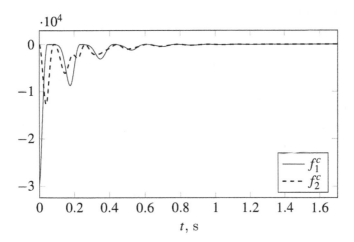

Figure 6.6: f_i^c's values during the control process.

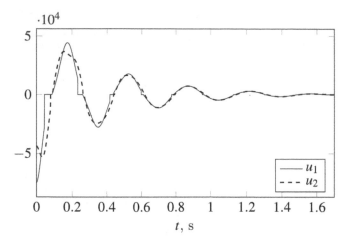

Figure 6.7: Control trajectories.

6.3 Constrained Bilinear Biquadratic Regulator

The previous section describes a solution method that was formulated by necessary optimality conditions. The developed method requires solving a nonlinear two points boundary value problem and an algorithm for its realization was suggested in Fig. 6.4. However, even though this algorithm is useful for simple models, like that given in Example 6.2.1, it fails to solve CLQR problems with more complex models. In this section, Krotov's method is used for solving a more general CLQR problem and it is proved to converge even for complex models. Namely, the CLQR problem is solved by addressing an equivalent problem, rather than treating it directly. First, a new CLQR problem is defined. Next, an equivalent problem, denoted as *constrained bilinear biquadratic regulator* (CBBR), is formulated. Finally, a novel sequence of improving functions, which corresponds to the CBBR problem, is derived and Krotov's method is applied.

Let a control force be admissible for semi-active dampers if it complies with constraints C1, C2, C3 and its magnitude is bounded by some positive scalar $w_{i,max}$. Additionally, assume that the plant is subjected to an external excitation. It should be emphasized that the control force bound is due to common practical design considerations, such as damper limitations or unwanted local force effects, rather than the damper's dissipative nature. These assumptions lead to following CLQR problem.

Definition 6.2 Let $\mathbf{x} : \mathbb{R} \to \mathbb{R}^n$ be a state trajectory and $\mathbf{w} : \mathbb{R} \to \mathbb{R}^{n_u}$ be a control forces trajectory. Let $\mathscr{W}_i(\mathbf{x})$ be a set of admissible control forces, where each $w_i \in \mathscr{W}_i(\mathbf{x})$ agrees with the following constraints:

C1: $w_i(t)\mathbf{c}_i\mathbf{x}(t) \leq 0$,

C2: $\mathbf{c}_i\mathbf{x}(t) = 0 \to w(t) = 0$,

C3.1: $|w_i(t)| \in [w_{i,min}(t,\mathbf{x}(t)), w_{i,max}]$,

for all $t \in [0,t_f]$ and for some $\mathbf{c}_i^T \in \mathbb{R}^n$, $w_{i,max} \geq 0$ and $w_{i,min} : \mathbb{R} \to [0, w_{i,max}]$.

The pair $(\mathbf{x}, \mathbf{w})$ is an admissible process if it satisfies the LTI state equation:

$$\dot{\mathbf{x}}(t) = \mathbf{A}\mathbf{x}(t) + \mathbf{B}\mathbf{w}(t) + \mathbf{g}(t); \quad \mathbf{x}(0), \forall t \in (0,t_f) \tag{6.10}$$

and $w_i \in \mathscr{W}_i(\mathbf{x})$ for $i = 1,\ldots,n_u$. Here $\mathbf{A} \in \mathbb{R}^{n \times n}$; $\mathbf{B} \in \mathbb{R}^{n \times n_u}$; $\mathbf{g} : \mathbb{R} \to \mathbb{R}^n$ is a trajectory of known, bounded, external excitations.

The control force w_i was defined in the above formulation as a mapping $\mathbb{R} \to \mathbb{R}$, i.e., merely as a function of time. However, as it can be found in [94], an equivalent bilinear mapping with suitable control trajectory bounds can be used for w_i, as follows.

Let the i-th control force $\hat{w}_i : \mathbb{R} \times \mathbb{R}^n \to \mathbb{R}$, be:

$$\hat{w}_i(t, \mathbf{x}(t)) = -u_i(t)\mathbf{c}_i\mathbf{x}(t) \tag{6.11}$$

where u_i is a control signal that satisfies

$$u_i(t) \begin{cases} = 0 & , \mathbf{c}_i\mathbf{x}(t) = 0 \\ \in \left[\frac{w_{i,min}(t,\mathbf{x}(t))}{|\mathbf{c}_i\mathbf{x}(t)|}, \frac{w_{i,max}}{|\mathbf{c}_i\mathbf{x}(t)|} \right] & , \text{otherwise} \end{cases} \tag{6.12}$$

The compliance of $\hat{w}_i$ with C2 is straightforward from Eq. (6.11). Therefore, the issue that was addressed by Lemmas 6.1 and 6.2 is avoided. It also follows that

$$\hat{w}_i(t, \mathbf{x}(t))\mathbf{c}_i\mathbf{x}(t) = -u_i(t)(\mathbf{c}_i\mathbf{x}(t))^2 \leq 0 \tag{6.13}$$

i.e., C1 is satisfied and

$$|\hat{w}_i(t,\mathbf{x}(t))| = |u_i(t)\mathbf{c}_i\mathbf{x}(t)| = u_i(t)|\mathbf{c}_i\mathbf{x}(t)| \leq \left(\frac{w_{i,max}}{|\mathbf{c}_i\mathbf{x}(t)|} \right) |\mathbf{c}_i\mathbf{x}(t)| = w_{i,max}$$

$$|\hat{w}_i(t,\mathbf{x}(t))| = u_i(t)|\mathbf{c}_i\mathbf{x}(t)| \geq \left(\frac{w_{i,min}(t,\mathbf{x}(t))}{|\mathbf{c}_i\mathbf{x}(t)|} \right) |\mathbf{c}_i\mathbf{x}(t)| = w_{i,min}(t,\mathbf{x}(t))$$

assures that C3.1 is satisfied too. Hence, the range of $\hat{w}_i$ coincides with the range of the functions in $\mathscr{W}_i(\mathbf{x})$, for any state trajectory $\mathbf{x}$. Substituting $\hat{\mathbf{w}}$ into Eq. (6.10), instead of $\mathbf{w}$, leads to an admissible process definition by means of a bilinear state-space equation.

Definition 6.3 Let $\mathbf{x} : \mathbb{R} \to \mathbb{R}^n$ be a state trajectory and $\mathbf{u} : \mathbb{R} \to \mathbb{R}^{n_u}$ be a control trajectory. Let $\mathscr{U}_i(\mathbf{x})$ be a set of admissible control trajectories where each $u \in \mathscr{U}_i(\mathbf{x})$ agrees with Eq. (6.12) for all $t \in [0, t_f]$ and for some $\mathbf{c}_i^T \in \mathbb{R}^n$, $w_{i,max} \geq 0$ and $w_{i,min} : \mathbb{R} \to [0, w_{i,max}]$.

The pair $(\mathbf{x}, \mathbf{u})$ is an admissible process if it satisfies the bilinear state equation

$$\dot{\mathbf{x}}(t) = \left(\mathbf{A} - \sum_{i=1}^{n_u} \mathbf{b}_i u_i(t) \mathbf{c}_i \right) \mathbf{x}(t) + \mathbf{g}(t); \quad \mathbf{x}(0), \forall t \in (0, t_f) \qquad (6.14)$$

and $u_i \in \mathscr{U}_i(\mathbf{x})$ for $i = 1, \ldots, n_u$. Here $\mathbf{A} \in \mathbb{R}^{n \times n}$; $\mathbf{g} : \mathbb{R} \to \mathbb{R}^n$ is a trajectory of known external excitations and $\mathbf{b}_i \in \mathbb{R}^n$.

The dependency of $\mathscr{U}_i$ on $\mathbf{x}$ is clear from Eq. (6.12). Let a performance index be

$$J(\mathbf{x}, \mathbf{w}) = \frac{1}{2} \int_0^{t_f} \left(\mathbf{x}(t)^T \mathbf{Q} \mathbf{x}(t) + \sum_{i=1}^{n_u} \left(w_i(t)^2 r_i + 2 w_i(t) \mathbf{n}_i \mathbf{x}(t) \right) \right) dt \qquad (6.15)$$

where $0 \leq \mathbf{Q} \in \mathbb{R}^{n \times n}$; $r_i > 0$ and $\mathbf{n}_i^T \in \mathbb{R}^n$ for $i = 1, \ldots, n_u$. Such form can be found in many practical control problems, where the control design is evaluated with respect to some linear output $\mathbf{y}$ (as in Eq. (2.6)). Substituting $\hat{\mathbf{w}}$ into Eq. (6.15), instead of $\mathbf{w}$, leads to the next optimal control problem.

Problem 6.2 (CBBR) *The constrained bilinear biquadratic regulator (CBBR) control problem is a search after an optimal and admissible process* $(\mathbf{x}^*, \mathbf{u}^*)$ *that minimizes the biquadratic performance index*

$$J(\mathbf{x}, \mathbf{u}) = \frac{1}{2} \int_0^{t_f} \left(\mathbf{x}(t)^T \mathbf{Q} \mathbf{x}(t) + \sum_{i=1}^{n_u} \left((\mathbf{c}_i \mathbf{x}(t))^2 u_i(t)^2 r_i - 2 u_i(t) (\mathbf{c}_i \mathbf{x}(t)) (\mathbf{n}_i \mathbf{x}(t)) \right) \right) dt$$

$$(6.16)$$

where $0 \leq \mathbf{Q} \in \mathbb{R}^{n \times n}$; $r_i > 0$ *and* $\mathbf{n}_i^T \in \mathbb{R}^n$ *for* $i = 1, \ldots, n_u$.

Krotov's method will be used for solving this problem. The following lemmas formulate a type of an improving function that suits the CBBR problem and enables using this method for solving it.

In the following derivations, some sums are replaced by matrix representations, as follows:

$$\sum_{i=1}^{n_u} u_i(t)^2 r_i(\mathbf{c}_i\mathbf{x}(t))^2 = \mathbf{x}(t)^T \mathbf{C}^T \, \mathbf{diag}((u_i(t)^2 r_i)_{i=1}^{n_u}) \mathbf{C}\mathbf{x}(t) \qquad (6.17)$$

$$\sum_{i=1}^{n_u} u_i(t)(\mathbf{c}_i\mathbf{x}(t))(\mathbf{n}_i\mathbf{x}(t)) = \mathbf{x}(t)^T \mathbf{C}^T \, \mathbf{diag}(\mathbf{u}(t)) \mathbf{N}\mathbf{x}(t) \qquad (6.18)$$

$$\sum_{i=1}^{n_u} \mathbf{b}_i u_i(t)\mathbf{c}_i\mathbf{x}(t) = \mathbf{B} \, \mathbf{diag}(\mathbf{u}(t)) \mathbf{C}\mathbf{x}(t) \qquad (6.19)$$

where $\mathbf{B} = \begin{bmatrix} \mathbf{b}_1 & \mathbf{b}_2 & ... \end{bmatrix}$, $\mathbf{N} = \begin{bmatrix} \mathbf{n}_1^T & \mathbf{n}_2^T & ... \end{bmatrix}^T$ and $\mathbf{C} = \begin{bmatrix} \mathbf{c}_1^T & \mathbf{c}_2^T & ... \end{bmatrix}^T$. $\mathscr{U}_i(t,\mathbf{x}) \subseteq \mathbb{R}^{n_u}$ is an intersection of $\mathscr{U}_i(\mathbf{x})$ at t, i.e., a set of control trajectories $\mathscr{U}_i(\mathbf{x})$, evaluated at a given t.

Lemma 6.4
Let

$$q(t,\xi) = \frac{1}{2}\xi^T \mathbf{P}(t)\xi + \mathbf{p}(t)^T \xi; \quad \mathbf{P}(t_f) = 0; \quad \mathbf{p}(t_f) = \mathbf{0}$$

where $\xi \in \mathbb{R}^n$, $\mathbf{P}: \mathbb{R} \to \mathbb{R}^{n \times n}$ is a continuous, piecewise differentiable and symmetric, matrix function and $\mathbf{p}: \mathbb{R} \to \mathbb{R}^n$ is a continuous and piecewise differentiable trajectory.

Let $v_i(t,\xi) \triangleq \frac{\xi^T(\mathbf{P}(t)\mathbf{b}_i + \mathbf{n}_i^T) + \mathbf{p}(t)^T \mathbf{b}_i}{r_i}$. The i-th element in the vector of control laws, $(\hat{u}_i)_{i=1}^{n_u}$, which minimizes $s(t,\mathbf{x}(t),\mathbf{u}(t))$, is

$$\hat{u}_i(t,\mathbf{x}(t)) = \arg \min_{\mathbf{v} \in \mathscr{U}_i(t,\mathbf{x})} s(t,\mathbf{x}(t),\mathbf{v})$$

$$= \begin{cases} 0 & , \mathbf{c}_i\mathbf{x}(t) = 0 \\ \dfrac{w_{i,min}(t,\mathbf{x}(t))}{|\mathbf{c}_i\mathbf{x}(t)|} & , \mathbf{c}_i\mathbf{x}(t) \neq 0, \, v_i(t,\mathbf{x}(t))\mathrm{sign}(\mathbf{c}_i\mathbf{x}(t)) \leq w_{i,min}(t,\mathbf{x}(t)) \\ \dfrac{w_{i,max}}{|\mathbf{c}_i\mathbf{x}(t)|} & , \mathbf{c}_i\mathbf{x}(t) \neq 0, \, v_i(t,\mathbf{x}(t))\mathrm{sign}(\mathbf{c}_i\mathbf{x}(t)) \geq w_{i,max} \\ \dfrac{v_i(t,\mathbf{x}(t))}{\mathbf{c}_i\mathbf{x}(t)} & , otherwise \end{cases}$$

$$(6.20)$$

Proof 6.5 The partial derivatives of q are:

$$q_t(t,\xi) = \frac{1}{2}\xi^T\dot{P}(t)\xi + \dot{p}(t)^T\xi \tag{6.21}$$

$$q_x(t,\xi) = \xi^T P(t) + p(t)^T \tag{6.22}$$

Let $v \in \mathbb{R}^{n_u}$. Substituting q_t and q_x into Eq. (3.18) and (3.19) yields:

$$s_f(x(t_f)) = 0 \quad \forall x(t_f) \in \mathscr{X}(t_f) \tag{6.23}$$

$$\begin{aligned}
s(t,x(t),v) &= q_t(t,x(t)) + q_x(t,x(t))f(t,x(t),v) \\
&\quad + \frac{1}{2}\left(x(t)^T Q x(t) + x(t)^T C^T \operatorname{diag}\left((v_i^2 r_i)_{i=1}^{n_u}\right) C x(t)\right. \\
&\quad \left. - 2x(t)^T C^T \operatorname{diag}(v) N x(t)\right)
\end{aligned} \tag{6.24}$$

$$\begin{aligned}
&= \frac{1}{2}x(t)^T \dot{P}(t)x(t) + \dot{p}(t)^T x(t) + \left(x(t)^T P(t) + p(t)^T\right) f(t,x(t),v) \\
&\quad + \frac{1}{2}x(t)^T Q x(t) + \frac{1}{2}x(t)^T C^T \operatorname{diag}\left((v_i^2 r_i)_{i=1}^{n_u}\right) C x(t) \\
&\quad - x(t)^T C^T \operatorname{diag}(v) N x(t)
\end{aligned} \tag{6.25}$$

$$\begin{aligned}
&= \frac{1}{2}x(t)^T \dot{P}(t)x(t) + \dot{p}(t)^T x(t) + x(t)^T P(t)\left(Ax(t) - B\operatorname{diag}(v)Cx(t)\right. \\
&\quad \left. + g(t)\right) + p(t)^T\left(Ax(t) - B\operatorname{diag}(v)Cx(t) + g(t)\right) + \frac{1}{2}x(t)^T Q x(t) \\
&\quad + \frac{1}{2}x(t)^T C^T \operatorname{diag}\left((v_i^2 r_i)_{i=1}^{n_u}\right) C x(t) - x(t)^T C^T \operatorname{diag}(v) N x(t)
\end{aligned} \tag{6.26}$$

$$\begin{aligned}
&= \frac{1}{2}x(t)^T \left(\dot{P}(t) + P(t)\left(A - B\operatorname{diag}(v)C\right) + \left(A - B\operatorname{diag}(v)C\right)^T P(t)\right. \\
&\quad \left. - C^T \operatorname{diag}(v)N - N^T \operatorname{diag}(v)C + C^T \operatorname{diag}\left((v_i^2 r_i)_{i=1}^{n_u}\right)C + Q\right)x(t) \\
&\quad + x(t)^T \left(\dot{p}(t) + \left(A - B\operatorname{diag}(v)C\right)^T p(t) + P(t)g(t)\right) + p(t)^T g(t)
\end{aligned} \tag{6.27}$$

$$\begin{aligned}
&= \frac{1}{2}x(t)^T \left(\dot{P}(t) + P(t)A + A^T P(t) + Q\right)x(t) + x(t)^T \left(\dot{p}(t) + A^T p(t) + P(t)g(t)\right) \\
&\quad + p(t)^T g(t) + \frac{1}{2}x(t)^T C^T \operatorname{diag}\left((v_i^2 r_i)_{i=1}^{n_u}\right)C x(t) \\
&\quad - x(t)^T \left(P(t)B + N^T\right)\operatorname{diag}(v)Cx(t) - p(t)^T B\operatorname{diag}(v)Cx(t)
\end{aligned} \tag{6.28}$$

$$\begin{aligned}
&= \frac{1}{2}x(t)^T \left(\dot{P}(t) + P(t)A + A^T P(t) + Q\right)x(t) \\
&\quad + x(t)^T \left(\dot{p}(t) + A^T p(t) + P(t)g(t)\right) + p(t)^T g(t) \\
&\quad + \frac{1}{2}\sum_{i=1}^{n_u}\left(r_i(c_i x(t))^2 v_i^2 - 2\left(x(t)^T(P(t)b_i + n_i^T) + p(t)^T b_i\right)c_i x(t) v_i\right)
\end{aligned} \tag{6.29}$$

$$=\frac{1}{2}\mathbf{x}(t)^T\left(\dot{\mathbf{P}}(t)+\mathbf{P}(t)\mathbf{A}+\mathbf{A}^T\mathbf{P}(t)+\mathbf{Q}\right)\mathbf{x}(t)$$
$$+\mathbf{x}(t)^T\left(\dot{\mathbf{p}}(t)+\mathbf{A}^T\mathbf{p}(t)+\mathbf{P}(t)\mathbf{g}(t)\right)+\mathbf{p}(t)^T\mathbf{g}(t) \qquad (6.30)$$
$$+\frac{1}{2}\sum_{i=1}^{n_u}\left(r_i(\mathbf{c}_i\mathbf{x}(t))^2 v_i^2 - 2r_i v_i(t,\mathbf{x}(t))\mathbf{c}_i\mathbf{x}(t)v_i\right)$$

where v_i was defined in the lemma. Completing the square leads to

$$s(t,\mathbf{x}(t),\mathbf{v})=\frac{1}{2}\mathbf{x}(t)^T\left(\dot{\mathbf{P}}(t)+\mathbf{P}(t)\mathbf{A}+\mathbf{A}^T\mathbf{P}(t)+\mathbf{Q}\right)\mathbf{x}(t)$$
$$+\mathbf{x}(t)^T\left(\dot{\mathbf{p}}(t)+\mathbf{A}^T\mathbf{p}(t)+\mathbf{P}(t)\mathbf{g}(t)\right)$$
$$+\mathbf{p}(t)^T\mathbf{g}(t)+\frac{1}{2}\sum_{i=1}^{n_u}\left(r_i\left(v_i\mathbf{c}_i\mathbf{x}(t)-v_i(t,\mathbf{x}(t))\right)^2 - r_i v_i(t,\mathbf{x}(t))^2\right)$$
$$=f_2(t,\mathbf{x}(t))+\frac{1}{2}\sum_{i=1}^{n_u}r_i\left(v_i\mathbf{c}_i\mathbf{x}(t)-v_i(t,\mathbf{x}(t))\right)^2$$

where $f_2:\mathbb{R}\times\mathbb{R}^n\to\mathbb{R}$ is some mapping independent of v_i. It follows that a minimum of $s(t,\mathbf{x}(t),\mathbf{v})$ over $\{\mathbf{v}|\mathbf{v}\in\mathscr{U}(t,\mathbf{x})\}$ is the minimum of the quadratic sum with respect to each $\{v_i|v_i\in\mathscr{U}_i(t,\mathbf{x})\}$, independently. Hence, the minimizing and admissible $v_i\in\mathscr{U}_i(t,\mathbf{x})$ is calculated as follows:

(a) When $\mathbf{c}_i\mathbf{x}(t)=0$, v_i vanishes from the performance index and the state equation. Hence its value has no effect and it can be set to $v_i=0$.

(b) When $\mathbf{c}_i\mathbf{x}(t)\neq0$ and $\frac{v_i(t,\mathbf{x}(t))}{\mathbf{c}_i\mathbf{x}(t)}\leq\frac{w_{i,min}(t,\mathbf{x}(t))}{|\mathbf{c}_i\mathbf{x}(t)|}$, the admissible minimum is attained at $v_i=\frac{w_{i,min}(t,\mathbf{x}(t))}{|\mathbf{c}_i\mathbf{x}(t)|}$. However, since

$$\frac{v_i(t,\mathbf{x}(t))}{\mathbf{c}_i\mathbf{x}(t)}\leq\frac{w_{i,min}(t,\mathbf{x}(t))}{|\mathbf{c}_i\mathbf{x}(t)|}\to v_i(t,\mathbf{x}(t))\frac{|\mathbf{c}_i\mathbf{x}(t)|}{\mathbf{c}_i\mathbf{x}(t)}\leq w_{i,min}(t,\mathbf{x}(t))$$
$$\to v_i(t,\mathbf{x}(t))\mathrm{sign}(\mathbf{c}_i\mathbf{x}(t))\leq w_{i,min}(t,\mathbf{x}(t))$$

it is possible to replace $\frac{v_i(t,\mathbf{x}(t))}{\mathbf{c}_i\mathbf{x}(t)}\leq\frac{w_{i,min}(t,\mathbf{x}(t))}{|\mathbf{c}_i\mathbf{x}(t)|}$ by $v_i(t,\mathbf{x}(t))\mathrm{sign}(\mathbf{c}_i\mathbf{x}(t))\leq w_{i,min}(t,\mathbf{x}(t))$.

(c) When $\mathbf{c}_i\mathbf{x}(t)\neq0$ and $\frac{v_i(t,\mathbf{x}(t))}{\mathbf{c}_i\mathbf{x}(t)}\geq\frac{w_{i,max}}{|\mathbf{c}_i\mathbf{x}(t)|}$, the admissible minimum is attained at $v_i=\frac{w_{i,max}}{|\mathbf{c}_i\mathbf{x}(t)|}$. However, since

$$\frac{v_i(t,\mathbf{x}(t))}{\mathbf{c}_i\mathbf{x}(t)}\geq\frac{w_{i,max}}{|\mathbf{c}_i\mathbf{x}(t)|}\to v_i(t,\mathbf{x}(t))\frac{|\mathbf{c}_i\mathbf{x}(t)|}{\mathbf{c}_i\mathbf{x}(t)}\geq w_{i,max}$$
$$\to v_i(t,\mathbf{x}(t))\mathrm{sign}(\mathbf{c}_i\mathbf{x}(t))\geq w_{i,max}$$

it is possible to replace $\frac{v_i(t,\mathbf{x}(t))}{\mathbf{c}_i\mathbf{x}(t)}\geq\frac{w_{i,max}}{|\mathbf{c}_i\mathbf{x}(t)|}$ by $v_i(t,\mathbf{x}(t))\mathrm{sign}(\mathbf{c}_i\mathbf{x}(t))\geq w_{i,max}$.

(d) When $\mathbf{c}_i\mathbf{x}(t) \neq 0$ and $\frac{w_{i,min}(t,\mathbf{x}(t))}{|\mathbf{c}_i\mathbf{x}(t)|} < \frac{v_i(t,\mathbf{x}(t))}{\mathbf{c}_i\mathbf{x}(t)} < \frac{w_{i,max}}{|\mathbf{c}_i\mathbf{x}(t)|}$, the admissible mini-
mum is attained at $v_i = \frac{v_i(t,\mathbf{x}(t))}{\mathbf{c}_i\mathbf{x}(t)}$.

The admissible minimizing $\hat{u}_i$ that corresponds to the admissible minimizing v_i is given by Eq. (6.20).

Lemma 6.5

Let $(\mathbf{x}_k, \mathbf{u}_k)$ be a given process, $\mathbf{P}_k$ and $\mathbf{p}_k$ be solutions to

$$\dot{\mathbf{P}}_k(t) = -\mathbf{P}_k(t)(\mathbf{A} - \mathbf{B}\,\mathbf{diag}(\mathbf{u}_k(t))\mathbf{C}) - (\mathbf{A} - \mathbf{B}\,\mathbf{diag}(\mathbf{u}_k(t))\mathbf{C})^T\mathbf{P}_k(t)$$
$$- \mathbf{Q} + \mathbf{C}^T\,\mathbf{diag}(\mathbf{u}_k(t))\mathbf{N} + \mathbf{N}^T\,\mathbf{diag}(\mathbf{u}_k(t))\mathbf{C} - \mathbf{C}^T\,\mathbf{diag}((u_{i,k}(t)^2 r_i)_{i=1}^{n_u})\mathbf{C}$$
$$\mathbf{P}_k(t_f) = \mathbf{0}$$

$$(6.31)$$

and

$$\dot{\mathbf{p}}_k(t) = -(\mathbf{A} - \mathbf{B}\,\mathbf{diag}(\mathbf{u}_k(t))\mathbf{C})^T\,\mathbf{p}_k(t) - \mathbf{P}_k(t)\mathbf{g}(t); \qquad \mathbf{p}_k(t_f) = \mathbf{0} \qquad (6.32)$$

Then

$$q_k(t,\mathbf{x}(t)) = \frac{1}{2}\mathbf{x}(t)^T\mathbf{P}_k(t)\mathbf{x}(t) + \mathbf{p}_k(t)^T\mathbf{x}(t)$$

is an improving function and the related s_k satisfies

$$s_k(t,\mathbf{x}_k(t),\mathbf{u}_k(t)) = \max_{\boldsymbol{\xi} \in \mathscr{X}(t)} s_k(t,\boldsymbol{\xi},\mathbf{u}_k(t)) \qquad (6.33)$$

Proof 6.6 According to Eq. (6.27), $s_k(t,\mathbf{x}(t),\mathbf{u}_k(t))$:

$$s_k(t,\mathbf{x}(t),\mathbf{u}_k(t)) = \frac{1}{2}\mathbf{x}(t)^T\Big(\dot{\mathbf{P}}_k(t) + \mathbf{P}_k(t)(\mathbf{A} - \mathbf{B}\,\mathbf{diag}(\mathbf{u}_k(t))\mathbf{C})$$
$$+ (\mathbf{A} - \mathbf{B}\,\mathbf{diag}(\mathbf{u}_k(t))\mathbf{C})^T\,\mathbf{P}_k(t) + \mathbf{Q} + \mathbf{C}^T\,\mathbf{diag}((u_{i,k}(t)^2 r_i)_{i=1}^{n_u})\mathbf{C}$$
$$- \mathbf{C}^T\,\mathbf{diag}(\mathbf{u}_k(t))\mathbf{N} - \mathbf{N}^T\,\mathbf{diag}(\mathbf{u}_k(t))\mathbf{C}\Big)\mathbf{x}(t)$$
$$+ \mathbf{x}(t)^T\Big(\dot{\mathbf{p}}_k(t) + (\mathbf{A} - \mathbf{B}\,\mathbf{diag}(\mathbf{u}_k(t))\mathbf{C})^T\,\mathbf{p}_k(t) + \mathbf{P}_k(t)\mathbf{g}(t)\Big)$$
$$+ \mathbf{p}_k(t)^T\mathbf{g}(t)$$

As $\dot{\mathbf{P}}_k(t)$ and $\dot{\mathbf{p}}_k(t)$ satisfy Eqs. (6.31) and (6.32):

$$s_k(t,\mathbf{x}(t),\mathbf{u}_k(t)) = \frac{1}{2}\mathbf{x}(t)^T\mathbf{0}\mathbf{x}(t) + \mathbf{x}(t)^T\mathbf{0} + \mathbf{p}_k(t)^T\mathbf{g}(t)$$
$$= \mathbf{p}_k(t)^T\mathbf{g}(t) \qquad (6.34)$$

As $s_k(t,\mathbf{x}(t),\mathbf{u}_k(t)) = s_k(t,\mathbf{x}_k(t),\mathbf{u}_k(t))$ it follows that

$$s_k(t,\mathbf{x}(t),\mathbf{u}_k(t))) \leq s_k(t,\mathbf{x}_k(t),\mathbf{u}_k(t))$$

for all $\mathbf{x}(t)$.

Remarks.

- Equation 6.34 leads to an alternative approach for computing $J(\mathbf{x}_k, \mathbf{u}_k)$. As $s_{kf}(\mathbf{x}(t_f)) = 0$ it follows that

$$
\begin{aligned}
J(\mathbf{x}_k, \mathbf{u}_k) =& J_{eq,k}(\mathbf{x}_k, \mathbf{u}_k) = q_k(0, \mathbf{x}(0)) + \int_0^{t_f} s_k(t, \mathbf{x}_k(t), \mathbf{u}_k(t)) \mathrm{d}t \\
=& \frac{1}{2} \mathbf{x}_k(0)^T \mathbf{P}_k(0) \mathbf{x}_k(0) + \mathbf{p}_k(0)^T \mathbf{x}_k(0) + \int_0^{t_f} \mathbf{p}_k(t)^T \mathbf{g}(t) \mathrm{d}t \quad (6.35)
\end{aligned}
$$

  This result will be generalized below in Lemma 6.9.

- When $\mathbf{g} = \mathbf{0}$ it follows that $\mathbf{p}_k(t) = \mathbf{0}$ and the problem is reduced to a free vibrations case.

- When there is no need for an upper bound for w_i, the same method can be used by letting $w_{i,max} = +\infty$.

Putting together the steps, described at the end of Section 4.3, and these two lemmas, allows computing two sequences, $\{q_k\}$ and $\{(\mathbf{x}_k, \mathbf{u}_k)\}$, such that the second one is an improving sequence. As J is non negative, it has an infimum and $\{(\mathbf{x}_k, \mathbf{u}_k)\}$ gets arbitrarily close to a candidate optimum.

The obtained algorithm is summarized in Fig. 6.8. Its output is an arbitrary approximation for $\mathbf{P}^*$ and $\mathbf{p}^*$, which defines the obtained control law (Eq. (6.20)). It should be noted that ostensibly the use of an absolute value in step (4) of the iterations is theoretically not necessary. However, it is still needed, as numerical computation errors might cause the algorithm to lose its monotonicity when it gets closer to an optimum.

In order that the suggested q will be proper, it should be smooth above $\mathcal{X}$, as well as continuous and piecewise differentiable above $(0, t_f)$. Its smoothness at any $\mathbf{x}(t) \in \mathcal{X}$ is obvious. However, its continuity and piecewise differentiability above $(0, t_f)$ is questionable, as follows. The definition of $\hat{u}_{i,k}$, consists of division in $\mathbf{c}_i \mathbf{x}$. Therefore, whenever $\mathbf{c}_i \mathbf{x}$ tends to zero, $\hat{u}_{i,k}$ tends to infinity. This issue does not pose a problem when solving the closed loop state equation, because the definition of $\hat{u}_{i,k}$ assures that $w_{i,k}$ will remain finite whenever $\mathbf{c}_i \mathbf{x}$ vanishes. However, the equations for $\mathbf{P}_k$ and $\mathbf{p}_k$ depend on $u_{i,k}$, rather than $\hat{u}_{i,k}$. Hence it might have discontinuities at any t that satisfies $\mathbf{c}_i \mathbf{x}(t) = 0$. This issue is addressed in the following lemmas.

Input:

$\mathbf{A}, \mathbf{B} = \begin{bmatrix} \mathbf{b}_1 & \mathbf{b}_2 & \ldots \end{bmatrix}$, $\mathbf{C} = \begin{bmatrix} \mathbf{c}_1^T & \mathbf{c}_2^T & \ldots \end{bmatrix}^T$, $\mathbf{g}$, $(w_{i,max})_{i=1}^{n_u}$, $(w_{i,min})_{i=1}^{n_u}$, $\mathbf{x}(0)$, $\mathbf{Q} \geq 0$, $(r_i | r_i > 0)_{i=1}^{n_u}$, $\mathbf{N} = \begin{bmatrix} \mathbf{n}_1^T & \mathbf{n}_2^T & \ldots \end{bmatrix}^T$.

Initialization:

(1) Select a convergence threshold - $\varepsilon > 0$.

(2) Solve

$$\dot{\mathbf{x}}_0(t) = (\mathbf{A} - \mathbf{B}\,\mathrm{diag}(\hat{\mathbf{u}}_0(t, \mathbf{x}_0(t))) \mathbf{C}) \mathbf{x}_0(t) + \mathbf{g}(t); \quad \mathbf{x}(0)$$

where

$$\hat{u}_{i,0}(t, \mathbf{x}(t)) = \begin{cases} 0 & , \mathbf{c}_i \mathbf{x}(t) = 0 \\ \dfrac{w_{i,min}(t, \mathbf{x}(t))}{|\mathbf{c}_i \mathbf{x}(t)|} & , \text{otherwise} \end{cases}$$

and set $\mathbf{u}_0(t) = \hat{\mathbf{u}}_0(t, \mathbf{x}_0(t))$. Solve

$$\dot{\mathbf{P}}_0(t) = -\mathbf{P}_0(t)(\mathbf{A} - \mathbf{B}\,\mathrm{diag}(\mathbf{u}_0(t))\mathbf{C}) - (\mathbf{A} - \mathbf{B}\,\mathrm{diag}(\mathbf{u}_0(t))\mathbf{C})^T \mathbf{P}_0(t)$$
$$- \mathbf{Q} + \mathbf{C}^T \,\mathrm{diag}(\mathbf{u}_0(t))\mathbf{N} + \mathbf{N}^T \,\mathrm{diag}(\mathbf{u}_0(t))\mathbf{C} - \mathbf{C}^T \,\mathrm{diag}((u_{i,0}(t)^2 r_i)_{i=1}^{n_u})\mathbf{C}$$
$$\dot{\mathbf{p}}_0(t) = -(\mathbf{A} - \mathbf{B}\,\mathrm{diag}(\mathbf{u}_0(t))\mathbf{C})^T \mathbf{p}_0(t) - \mathbf{P}_0(t)\mathbf{g}(t)$$

for $\mathbf{P}_0(t_f) = \mathbf{0}$ and $\mathbf{p}_0(t_f) = \mathbf{0}$.

(3) Set $w_{i,0}(t) = -u_{i,0}(t)\mathbf{c}_i \mathbf{x}_0(t)$ and compute

$$J(\mathbf{x}_0, \mathbf{w}_0) = \frac{1}{2} \int_0^{t_f} \mathbf{x}_0(t)^T \mathbf{Q}\mathbf{x}_0(t) + \sum_{i=1}^{n_u} w_{i,0}(t)^2 r_i + 2w_{i,0}(t)\mathbf{n}_i \mathbf{x}_0(t)\, dt$$

Iterations: For $k = \{0, 1, 2, \ldots\}$:

(1) Propagate to the improved process by solving

$$\dot{\mathbf{x}}_{k+1}(t) = (\mathbf{A} - \mathbf{B}\,\mathrm{diag}(\hat{\mathbf{u}}_{k+1}(t, \mathbf{x}_{k+1}(t)))\mathbf{C}) \mathbf{x}_{k+1}(t) + \mathbf{g}(t); \quad \mathbf{x}_{k+1}(0) = \mathbf{x}(0) \quad (6.36)$$

where $v_{i,k}(t, \mathbf{x}(t)) \triangleq \left(\left(\mathbf{b}_i^T \mathbf{P}_k(t) + \mathbf{n}_i \right) \mathbf{x}(t) + \mathbf{b}_i^T \mathbf{p}_k(t) \right) / r_i$

Figure 6.8: CBBR—Algorithm for successive improvement of control process.

and

$$
\hat{u}_{i,k+1}(t,\mathbf{x}(t)) = \begin{cases} 0 & ,\mathbf{c}_i\mathbf{x}(t)=0 \\[2mm] \dfrac{w_{i,min}(t,\mathbf{x}(t))}{|\mathbf{c}_i\mathbf{x}(t)|} & ,\mathbf{c}_i\mathbf{x}(t)\neq 0,\, v_{i,k}(t,\mathbf{x}(t))\mathrm{sign}(\mathbf{c}_i\mathbf{x}(t)) \leq w_{i,min}(t,\mathbf{x}(t)) \\[2mm] \dfrac{w_{i,max}}{|\mathbf{c}_i\mathbf{x}(t)|} & ,\mathbf{c}_i\mathbf{x}(t)\neq 0,\, v_{i,k}(t,\mathbf{x}(t))\mathrm{sign}(\mathbf{c}_i\mathbf{x}(t)) \geq w_{i,max} \\[2mm] \dfrac{v_{i,k}(t,\mathbf{x}(t)))}{\mathbf{c}_i\mathbf{x}(t)} & ,\text{otherwise} \end{cases}
$$

Set $\mathbf{u}_{k+1}(t) = \hat{\mathbf{u}}_{k+1}(t,\mathbf{x}_{k+1}(t))$.

(2) Solve

$$
\dot{\mathbf{P}}_{k+1}(t) = -\mathbf{P}_{k+1}(t)(\mathbf{A}-\mathbf{B}\,\mathrm{diag}(\mathbf{u}_{k+1}(t))\mathbf{C}) - (\mathbf{A}-\mathbf{B}\,\mathrm{diag}(\mathbf{u}_{k+1}(t))\mathbf{C})^T\mathbf{P}_{k+1}(t)
$$
$$
-\,\mathbf{Q}+\mathbf{C}^T\,\mathrm{diag}(\mathbf{u}_{k+1}(t))\mathbf{N}+\mathbf{N}^T\,\mathrm{diag}(\mathbf{u}_{k+1}(t))\mathbf{C}-\mathbf{C}^T\,\mathrm{diag}((u_{i,k+1}(t)^2 r_i)_{i=1}^{n_u})\mathbf{C}
$$

$$
\dot{\mathbf{p}}_{k+1}(t) = -(\mathbf{A}-\mathbf{B}\,\mathrm{diag}(\mathbf{u}_{k+1}(t))\mathbf{C})^T\,\mathbf{p}_{k+1}(t)-\mathbf{P}_{k+1}(t)\mathbf{g}(t)
$$

for $\mathbf{P}_{k+1}(t_f)=\mathbf{0}$ and $\mathbf{p}_{k+1}(t_f)=\mathbf{0}$.

(3) Set $w_{i,k+1}(t) = -u_{i,k+1}(t)\mathbf{c}_i\mathbf{x}_{k+1}(t)$ and compute

$$
J(\mathbf{x}_{k+1},\mathbf{w}_{k+1}) = \frac{1}{2}\int_0^{t_f} \mathbf{x}_{k+1}(t)^T\mathbf{Q}\mathbf{x}_{k+1}(t)
$$
$$
+\sum_{i=1}^{n_u} w_{i,k+1}(t)^2 r_i + 2w_{i,k+1}(t)\mathbf{n}_i\mathbf{x}_{k+1}(t)\mathrm{d}t
$$

(4) If $|J(\mathbf{x}_k,\mathbf{u}_k)-J(\mathbf{x}_{k+1},\mathbf{u}_{k+1})| < \varepsilon$, stop iterating, otherwise - continue.

Output: $\mathbf{P}_{k+1},\,\mathbf{p}_{k+1}$.

Figure 6.8(Cont.): CBBR—algorithm for successive improvement of control process.

Let $C(I)$ be the set of continuous functions $I \to \mathbb{R}$.

Lemma 6.6
Let linear matrix and vector ODEs be

$$
\dot{\mathbf{P}}(t) = -\mathbf{P}(t)\mathbf{A}(t)-\mathbf{A}(t)^T\mathbf{P}(t)-\mathbf{Q}(t); \quad \forall t \in T; \quad \mathbf{P}(t_0) \tag{6.37a}
$$
$$
\dot{\mathbf{p}}(t) = -\mathbf{A}(t)^T\mathbf{p}(t)-\mathbf{P}(t)\mathbf{g}(t); \quad \forall t \in T; \quad \mathbf{p}(t_0) \tag{6.37b}
$$

where $T \subset \mathbb{R}$ is an open interval, $t_0 \in T$; $\mathbf{A}, \mathbf{Q} : T \to \mathbb{R}^{n \times n}$ are matrix functions and $\mathbf{g} : T \to \mathbb{R}^n$. $\mathbf{P}(t_0)$ and $\mathbf{p}(t_0)$ are given value conditions. If all the elements of $\mathbf{A}$, $\mathbf{Q}$ and $\mathbf{g}$ are members of $C(T)$ then there exist unique $\mathbf{P}$ and $\mathbf{p}$, whose elements are smooth.

Proof 6.7 Let the hypothesis hold and let

$$\mathbf{F}(t, \mathbf{P}(t)) \triangleq -\mathbf{P}(t)\mathbf{A}(t) - \mathbf{A}(t)^T \mathbf{P}(t) - \mathbf{Q}(t)$$

Hence

$$\|\mathbf{F}(t, \mathbf{P}_a(t)) - \mathbf{F}(t, \mathbf{P}_b(t))\| = \|(\mathbf{P}_a(t) - \mathbf{P}_b(t))\mathbf{A}(t) + \mathbf{A}(t)^T (\mathbf{P}_a(t) - \mathbf{P}_b(t))\|$$
$$\leq 2\|\mathbf{A}(t)\|\|\mathbf{P}_a(t) - \mathbf{P}_b(t)\|$$

By continuity of $\mathbf{A}$'s elements, $\mathbf{A}(t)$ is a continuous linear operator at any $t \in T$. Therefore $\mathbf{A}(t)$ is a bounded linear operator at any $t \in T$. Hence

$$\|\mathbf{F}(t, \mathbf{P}_a(t)) - \mathbf{F}(t, \mathbf{P}_b(t))\| \leq L\|\mathbf{P}_a(t) - \mathbf{P}_b(t)\| \tag{6.38}$$

for all $t \in T$, where

$$L \triangleq 2\|\mathbf{A}(t)\|$$

As this also holds for any open sub-interval $I \subset T$, it follows that $\mathbf{F}$ is locally Lipschitz over T and a unique solution $\mathbf{P}$ exists for T. As every element of $\mathbf{Q}$ is a member of $C(T)$, $\mathbf{F}$ is continuous at any of its arguments and $\mathbf{P}$ is continuous [8]. It follows that the elements of $\dot{\mathbf{P}}(t) = \mathbf{F}(t, \mathbf{P}(t))$ are members of $C(T)$. Hence $\mathbf{P}$ is smooth in T. Similar steps can be carried out in order to prove $\mathbf{p}$'s smoothness.

Lemma 6.7
Let $(\mathbf{A})_{ij}$, $(\mathbf{Q})_{ij}$, g_i be elements of $\mathbf{A}$, $\mathbf{Q}$ and $\mathbf{g}$, and $D_{ij}^{\mathbf{A}}, D_{ij}^{\mathbf{Q}}, D_i^g \subset T$ be sets of time instances, when $(\mathbf{A})_{ij}$, $(\mathbf{Q})_{ij}$, g_i are discontinuous, respectively. If $D_{ij}^{\mathbf{A}}, D_{ij}^{\mathbf{Q}}, D_i^g$ are countable, then Eq. (6.37) have solutions $\mathbf{P}$, $\mathbf{p}$ that are continuous and piecewise differentiable above T.

Proof 6.8 Let

$$D \triangleq \left(\cup_{i,j \in [1,\ldots,n]} D_{ij}^{\mathbf{A}} \right) \cup \left(\cup_{i,j \in [1,\ldots,n]} D_{ij}^{\mathbf{Q}} \right) \tag{6.39}$$

be the set of all discontinuities in $\mathbf{A}$ and $\mathbf{Q}$. Let $I \triangleq T \setminus D$, that is, I is the set of all $t \in T$ in which all the elements of $\mathbf{A}$ and $\mathbf{Q}$ are continuous. Let $I_0, I_1, I_2, \ldots \subset I$ be ordered, separated sub-intervals of I such that $I = \cup I_m$. For each I_m define exactly one value condition $\mathbf{P}^m(t_m)$ at $t_m \in I_m$ such that if $t_M \in I_M$ for some M then $\mathbf{P}(t_M) = \mathbf{P}(t_0)$.

It follows that for each I_m, a value condition problem is defined by Eq. (6.37a) and the corresponding value condition $\mathbf{P}^m(t_m)$. As all elements of $\mathbf{A}$ and $\mathbf{Q}$ are in $C(I_m)$, Lemma 6.6 assures that there exists a unique and smooth solution $\mathbf{P}^m$ for I_m. By an appropriate selection of value conditions $\{\mathbf{P}^m(t_m)\}$ and ordering the set of solutions $\{\mathbf{P}^m\}$ 'head to tail', a continuous and piecewise differentiable solution $\mathbf{P}$ can be constructed such that the value condition $\mathbf{P}(t_0)$ is satisfied.

Let

$$\Delta \triangleq \left(\cup_{i,j \in [1,\ldots,n]} D_{ij}^{\mathbf{A}} \right) \cup \left(\cup_{i \in [1,\ldots,n]} D_i^{\mathbf{g}} \right) \tag{6.40}$$

be the set of all discontinuities in $\mathbf{A}$ and $\mathbf{g}$. By continuity of $\mathbf{P}$, it follows that $\mathbf{Pg}$ has the same discontinuities as $\mathbf{g}$. Let $\eta \triangleq T \setminus \Delta$, that is, η is a set of $t \in T$ in which all the elements of $\mathbf{A}$ and $\mathbf{Pg}$ are continuous. Let $\eta_0, \eta_1, \eta_2, \ldots \subset \eta$ be ordered and separated sub-intervals of η such that $\eta = \cup \eta_m$. For each η_m define exactly one value condition $\mathbf{p}^m(t_m)$ at $t_m \in \eta_m$ such that if $t_M \in \eta_M$ for some M then $\mathbf{p}(t_M) = \mathbf{p}(t_0)$.

It follows that for each η_m, a value condition problem is defined by Eq. (6.37a) and the corresponding value condition $\mathbf{p}^m(t_m)$. As all elements of $\mathbf{A}$ and $\mathbf{Pg}$ are members of $C(I_m)$, Lemma 6.6 assures that there exists a unique and smooth solution $\mathbf{p}^m$ for I_m. By an appropriate selection of value conditions $\{\mathbf{p}^m(t_m)\}$ and ordering the set of solutions $\{\mathbf{p}^m\}$ 'head to tail', a continuous and piecewise differentiable solution $\mathbf{p}$ can be constructed such that the value condition $\mathbf{p}(t_0)$ is satisfied.

If $\mathbf{u}_k$ and $\mathbf{g}$ have piecewise-continuous elements, then the same is true for $(\mathbf{A} - \mathbf{B}\,\mathbf{diag}(\mathbf{u}_{k+1})\mathbf{C})$;

$$(-\mathbf{Q} + \mathbf{C}^T \mathbf{diag}(\mathbf{u}_{k+1})\mathbf{N} + \mathbf{N}^T \mathbf{diag}(\mathbf{u}_{k+1})\mathbf{C} - \mathbf{C}^T \mathbf{diag}((u_{i,k+1}^2 r_i)_{i=1}^{n_u})\mathbf{C});$$

and $\mathbf{g}$. It follows from Lemma 6.7 that there exist corresponding $\mathbf{P}_k$ and $\mathbf{p}_k$ that are continuous and piecewise-differentiable above $(0, t_f)$. Consequently, q_k is continuous and piecewise-differentiable above $(0, t_f)$, i.e., it is a proper improving function.

The following paragraphs discuss further properties of $\mathbf{P}$ and $\mathbf{p}$ that solve Eqs. (6.31) and (6.32). Let $\mathbf{\Phi} : \mathbb{R} \times \mathbb{R} \to \mathbb{R}^{n \times n}$ be a transition matrix that satisfies $\dot{\mathbf{\Phi}}(t, \tau) = \mathbf{A}(t)\mathbf{\Phi}(t, \tau)$ for some $\mathbf{A} : \mathbb{R} \to \mathbb{R}^{n \times n}$.

The following identities will be useful in the subsequent derivations: (a) $\frac{d}{d\tau}\mathbf{\Phi}(t,\tau)^T = -\mathbf{A}(\tau)^T\mathbf{\Phi}(t,\tau)^T$, (b) $\mathbf{\Phi}(\tau,\tau) = \mathbf{I}$ [59].

The solution of $\dot{\mathbf{x}}(t) = \mathbf{A}(t)\mathbf{x}(t) + \mathbf{g}(t)$, with a given $\mathbf{x}(0)$, is $\mathbf{x}(t) = \mathbf{\Phi}(t,0)\mathbf{x}(0) + \int_0^t \mathbf{\Phi}(t,\tau)\mathbf{g}(\tau)d\tau$. Similarly, the solution of $\dot{\mathbf{p}}(t) = -\mathbf{A}(t)^T\mathbf{p}(t) - \mathbf{g}(t)$ for a given $\mathbf{p}(t_f)$ is $\mathbf{p}(t) = \mathbf{\Phi}(t_f,t)^T\mathbf{p}(t_f) + \int_t^{t_f} \mathbf{\Phi}(\tau,t)^T\mathbf{g}(\tau)d\tau$.

Let $\mathbf{P}$ satisfy $\dot{\mathbf{P}}(t) = -\mathbf{P}(t)\mathbf{A}(t) - \mathbf{A}(t)^T\mathbf{P}(t) - \mathbf{Q}(t)$ for a given $\mathbf{P}(t_f)$. It can be written explicitly by [33]:

$$\mathbf{P}(t) = \mathbf{\Phi}(t_f,t)^T\mathbf{P}(t_f)\mathbf{\Phi}(t_f,t) + \int_t^{t_f}\mathbf{\Phi}(\tau,t)^T\mathbf{Q}(\tau)\mathbf{\Phi}(\tau,t)d\tau \qquad (6.41)$$

It can be verified easily by differentiating both sides with respect to t:

$$\dot{\mathbf{P}}(t) = \dot{\mathbf{\Phi}}(t_f,t)^T\mathbf{P}(t_f)\mathbf{\Phi}(t_f,t) + \int_t^{t_f}\dot{\mathbf{\Phi}}(\tau,t)^T\mathbf{Q}(\tau)\mathbf{\Phi}(\tau,t)d\tau$$

$$+ \mathbf{\Phi}(t_f,t)^T\mathbf{P}(t_f)\dot{\mathbf{\Phi}}(t_f,t) + \int_t^{t_f}\mathbf{\Phi}(\tau,t)^T\mathbf{Q}(\tau)\dot{\mathbf{\Phi}}(\tau,t)d\tau$$

$$- \mathbf{\Phi}(t,t)^T\mathbf{Q}(t)\mathbf{\Phi}(t,t)$$

$$= -\mathbf{A}(t)^T\left(\mathbf{\Phi}(t_f,t)^T\mathbf{P}(t_f)\mathbf{\Phi}(t_f,t) + \int_t^{t_f}\mathbf{\Phi}(\tau,t)^T\mathbf{Q}(\tau)\mathbf{\Phi}(\tau,t)d\tau\right)$$

$$- \left(\mathbf{\Phi}(t_f,t)^T\mathbf{P}(t_f)\mathbf{\Phi}(t_f,t) + \int_t^{t_f}\mathbf{\Phi}(\tau,t)^T\mathbf{Q}(\tau)\mathbf{\Phi}(\tau,t)d\tau\right)\mathbf{A}(t) - \mathbf{Q}(t)$$

$$= -\mathbf{A}(t)^T\mathbf{P}(t) - \mathbf{P}(t)\mathbf{A}(t) - \mathbf{Q}(t)$$

The following lemmas address some properties of $\mathbf{P}$ and $\mathbf{p}$, and some meanings of J_{eq}'s components. These results allow to evaluate the accuracy of computed $\mathbf{P}_k$ and $\mathbf{p}_k$, as explained hereinafter.

Lemma 6.8

If

$$\begin{bmatrix} \mathbf{Q} & -\mathbf{N}^T \\ -\mathbf{N} & \mathbf{diag}((r_i)_{i=1}^{n_u}) \end{bmatrix} \geq 0 \tag{6.42}$$

then a $\mathbf{P}_k$ *that solves Eq.* (6.31) *satisfies* $\mathbf{P}_k(t) \geq 0$ *for all* $t \in [0, t_f]$ *and any control trajectory* $\mathbf{u}_k \in \{\mathbb{R} \to \mathbb{R}^{n_u}\}$.

Proof 6.9   Assume that Eq. (6.42) holds. Hence

$$
\begin{aligned}
\hat{\mathbf{Q}}(t) &\triangleq \mathbf{Q} + \mathbf{C}^T \, \mathbf{diag}((u_{i,k}(t)^2 r_i)_{i=1}^{n_u}) \mathbf{C} \\
&\quad - \mathbf{C}^T \, \mathbf{diag}(\mathbf{u}_k(t)) \mathbf{N} - \mathbf{N}^T \, \mathbf{diag}(\mathbf{u}_k(t)) \mathbf{C} \\
&= \begin{bmatrix} \mathbf{I} & \mathbf{C}^T \, \mathbf{diag}(\mathbf{u}_k(t)) \end{bmatrix} \begin{bmatrix} \mathbf{Q} & -\mathbf{N}^T \\ -\mathbf{N} & \mathbf{diag}((r_i)_{i=1}^{n_u}) \end{bmatrix} \begin{bmatrix} \mathbf{I} \\ \mathbf{diag}(\mathbf{u}_k(t)) \mathbf{C} \end{bmatrix} \geq 0
\end{aligned}
\tag{6.43}
$$

and

$$\mathbf{P}(t) = \int_t^{t_f} \mathbf{\Phi}(t,\tau)^T \tilde{\mathbf{Q}}(t) \mathbf{\Phi}(t,\tau) \mathrm{d}t \geq 0 \tag{6.44}$$

Lemma 6.9
Let $[t_1, t_2] \subseteq [0, t_f]$ *and define functionals* $\bar{J} : \mathscr{X} \times \mathscr{U} \times \mathbb{R}^2 \to \mathbb{R}$ *and* $\bar{J}_{eq} : \mathscr{X} \times \mathscr{U} \times \mathbb{R}^2 \to \mathbb{R}$ *as:*

$$\bar{J}(\mathbf{x}, \mathbf{u}, t_1, t_2) \triangleq \int_{t_1}^{t_2} l(t, \mathbf{x}(t), \mathbf{u}(t)) \mathrm{d}t \tag{6.45}$$

$$\bar{J}_{eq}(\mathbf{x}, \mathbf{u}, t_1, t_2) \triangleq q(t_1, \mathbf{x}(t_1)) - q(t_2, \mathbf{x}(t_2)) + \int_{t_1}^{t_2} s(t, \mathbf{x}(t), \mathbf{u}(t)) \mathrm{d}t \tag{6.46}$$

If $(\mathbf{x}, \mathbf{u})$ *is admissible, then* $\bar{J}(\mathbf{x}, \mathbf{u}, t_1, t_2) = \bar{J}_{eq}(\mathbf{x}, \mathbf{u}, t_1, t_2)$ *for any* $0 \leq t_1 \leq t_2 \leq t_f$.

Proof 6.10   By subtracting $\bar{J}$ from $\bar{J}_{eq}$ and using Eq. (3.18):

$$
\begin{aligned}
\bar{J}_{eq}(\mathbf{x}, \mathbf{u}, t_1, t_2) - \bar{J}(\mathbf{x}, \mathbf{u}, t_1, t_2) &= q(t_1, \mathbf{x}(t_1)) - q(t_2, \mathbf{x}(t_2)) \\
&\quad + \int_{t_1}^{t_2} (q_t(t, \mathbf{x}(t)) + q_{\mathbf{x}}(t, \mathbf{x}(t)) \mathbf{f}(t, \mathbf{x}(t), \mathbf{u}(t))) \, \mathrm{d}t
\end{aligned}
$$

As $(\mathbf{x}, \mathbf{u})$ is an admissible process:

$$\bar{J}_{eq}(\mathbf{x}, \mathbf{u}, t_1, t_2) - \bar{J}(\mathbf{x}, \mathbf{u}, t_1, t_2) = q(t_1, \mathbf{x}(t_1)) - q(t_2, \mathbf{x}(t_2))$$

$$+ \int_{t_1}^{t_2} (q_t(t, \mathbf{x}(t)) + q_{\mathbf{x}}(t, \mathbf{x}(t))\dot{\mathbf{x}}(t)) \, dt$$

$$= q(t_1, \mathbf{x}(t_1)) - q(t_2, \mathbf{x}(t_2)) + \int_{t_1}^{t_2} dq(t, \mathbf{x}(t)) = 0$$

by the virtue of Newton-Leibnitz formula.

Substituting q_k and s_k, defined in Lemma 6.5, into $\bar{J}_{eq}$, yields

$$\bar{J}_{eq}(\mathbf{x}, \mathbf{u}, t_1, t_2) = \frac{1}{2}\mathbf{x}_k(t_1)^T \mathbf{P}_k(t_1)\mathbf{x}_k(t_1) + \mathbf{p}_k(t_1)^T \mathbf{x}_k(t_1)$$

$$- \frac{1}{2}\mathbf{x}_k(t_2)^T \mathbf{P}_k(t_2)\mathbf{x}_k(t_2) - \mathbf{p}_k(t_2)^T \mathbf{x}_k(t_2) + \int_{t_1}^{t_2} \mathbf{p}_k(t)^T \mathbf{g}(t) dt \qquad (6.47)$$

and by letting $t_2 = t_f$:

$$\bar{J}_{eq}(\mathbf{x}, \mathbf{u}, t_1, t_f) = \frac{1}{2}\mathbf{x}_k(t_1)^T \mathbf{P}_k(t_1)\mathbf{x}_k(t_1) + \mathbf{p}_k(t_1)^T \mathbf{x}_k(t_1) + \int_{t_1}^{t_f} \mathbf{p}_k(t)^T \mathbf{g}(t) dt$$

$$(6.48)$$

If $t_1 = 0$, then $\bar{J}_{eq}(\mathbf{x}, \mathbf{u}, t_1, t_f)$ becomes Eq. (6.35). This simple result has a practical significance. Numerical observations show that the improvement from $(\mathbf{x}_k, \mathbf{u}_k)$ to $(\mathbf{x}_{k+1}, \mathbf{u}_{k+1})$, i.e., stage 1 in Fig. 6.8, is quite sensitive to the accuracy of $\mathbf{P}_k$ and $\mathbf{p}_k$ that solves Eqs. (6.31) and (6.32), i.e., stage 2 in Fig. 6.8. Lemma 6.9 provides a measure to that accuracy. As $\mathbf{P}_k(t_f)$ and $\mathbf{p}_k(t_f)$ are known accurately and as the computation of $\bar{J}(\mathbf{x}_k, \mathbf{u}_k, t_1, t_f)$ is independent of $\mathbf{P}_k$ and $\mathbf{p}_k$, one can evaluate the accuracy of $\mathbf{P}_k(t_1)$ and $\mathbf{p}_k(t_1)$ by examining the difference $|\bar{J}_{eq}(\mathbf{x}_k, \mathbf{u}_k, t_1, t_f) - \bar{J}(\mathbf{x}_k, \mathbf{u}_k, t_1, t_f)|$. Higher accuracy will lead to smaller difference. This is an important tool as numerical errors are always an issue in the numerical solution of differential equation and also because the differential Lyapunov equations tend to be quite stiff [60].

Let $(\mathbf{x}, \mathbf{u})$ be an admissible process in interval $[0, t_f]$. Let $\mathbf{x} = \mathbf{x}_h + \mathbf{x}_g$, where $\mathbf{x}_h$ is a solution to $\dot{\mathbf{x}}_h(t) = \mathbf{A}_c(t)\mathbf{x}_h(t)$ for $\mathbf{x}_h(0) = \mathbf{x}(0)$ and $\mathbf{x}_g$ is a solution of $\dot{\mathbf{x}}_g(t) = \mathbf{A}_c(t)\mathbf{x}_g(t) + \mathbf{g}(t)$ for $\mathbf{x}_g(0) = \mathbf{0}$. The performance index can be written as

$$J(\mathbf{x}_h + \mathbf{x}_g, \mathbf{u}) = \frac{1}{2} \int_0^{t_f} \mathbf{x}_h(t)^T \mathbf{Q} \mathbf{x}_h(t) \mathrm{d}t + \frac{1}{2} \int_0^{t_f} \mathbf{x}_g(t)^T \mathbf{Q} \mathbf{x}_g(t) \mathrm{d}t$$

$$+ \int_0^{t_f} \mathbf{x}_h(t)^T \mathbf{Q} \mathbf{x}_g(t) \mathrm{d}t + \frac{1}{2} \int_0^{t_f} \mathbf{x}_g(t)^T \mathbf{C}^T \, \mathbf{diag}((u_i(t)^2 r_i)_{i=1}^{n_u}) \mathbf{C} \mathbf{x}_g(t) \mathrm{d}t$$

$$+ \frac{1}{2} \int_0^{t_f} \mathbf{x}_h(t)^T \mathbf{C}^T \, \mathbf{diag}((u_i(t)^2 r_i)_{i=1}^{n_u}) \mathbf{C} \mathbf{x}_h(t) \mathrm{d}t$$

$$+ \int_0^{t_f} \mathbf{x}_h(t)^T \mathbf{C}^T \, \mathbf{diag}((u_i(t)^2 r_i)_{i=1}^{n_u}) \mathbf{C} \mathbf{x}_g(t) \mathrm{d}t$$

$$- \int_0^{t_f} \mathbf{x}_g(t)^T \mathbf{N}^T \, \mathbf{diag}(\mathbf{u}(t)) \mathbf{C} \mathbf{x}_g(t) \mathrm{d}t - \int_0^{t_f} \mathbf{x}_h(t)^T \mathbf{N}^T \, \mathbf{diag}(\mathbf{u}(t)) \mathbf{C} \mathbf{x}_h(t) \mathrm{d}t$$

$$- 2 \int_0^{t_f} \mathbf{x}_h(t)^T \mathbf{N}^T \, \mathbf{diag}(\mathbf{u}(t)) \mathbf{C} \mathbf{x}_g(t) \mathrm{d}t$$

The following two lemmas relate each of the elements in the above formulation with a corresponding one in J_{eq}.

$\boldsymbol{\Phi}$ is a state transition matrix that satisfies

$$\dot{\boldsymbol{\Phi}}(t, \tau) = (\mathbf{A} - \mathbf{B} \, \mathbf{diag}(\mathbf{u}(t)) \mathbf{C}) \boldsymbol{\Phi}(t, \tau)$$

with $\boldsymbol{\Phi}(\tau, \tau) = \mathbf{I}$.

Lemma 6.10
Let $\mathbf{A}_c(t) \triangleq \mathbf{A} - \mathbf{B} \, \mathbf{diag}(\mathbf{u}(t)) \mathbf{C}$. $\hat{\mathbf{Q}} : \mathbb{R} \to \mathbb{R}^{n \times n}$ *is a symmetric matrix function.*
If $\mathbf{x}_h, \mathbf{x}_g, \mathbf{p} \in \{\mathbb{R} \to \mathbb{R}^n\}$ *and* $\mathbf{P} \in \{\mathbb{R} \to \mathbb{R}^{n \times n}\}$ *satisfy the linear ODEs*

$$\begin{aligned}
\dot{\mathbf{x}}_h(t) &= \mathbf{A}_c(t) \mathbf{x}_h(t); & \mathbf{x}_h(t_1) &= \mathbf{x}(t_1) & \text{(6.49a)} \\
\dot{\mathbf{x}}_g(t) &= \mathbf{A}_c(t) \mathbf{x}_g(t) + \mathbf{g}(t); & \mathbf{x}_g(t_1) &= \mathbf{0} & \text{(6.49b)} \\
\dot{\mathbf{P}}(t) &= -\mathbf{P}(t) \mathbf{A}_c(t) - \mathbf{A}_c(t)^T \mathbf{P}(t) - \hat{\mathbf{Q}}(t); & \mathbf{P}(t_f) &= \mathbf{0} & \text{(6.49c)} \\
\dot{\mathbf{p}}(t) &= -\mathbf{A}_c(t)^T \mathbf{p}(t) - \mathbf{P}(t) \mathbf{g}(t); & \mathbf{p}(t_f) &= \mathbf{0} & \text{(6.49d)}
\end{aligned}$$

for $t \in (t_1, t_f)$, *Then:*

(a) $\frac{1}{2} \mathbf{x}(t_1)^T \mathbf{P}(t_1) \mathbf{x}(t_1) = \frac{1}{2} \int_{t_1}^{t_f} \mathbf{x}_h(t)^T \hat{\mathbf{Q}}(t) \mathbf{x}_h(t) \mathrm{d}t$

(b) $\int_{t_1}^{t_f} \mathbf{p}(t)^T \mathbf{g}(t) \mathrm{d}t = \frac{1}{2} \int_{t_1}^{t_f} \mathbf{x}_g(t)^T \hat{\mathbf{Q}}(t) \mathbf{x}_g(t) \mathrm{d}t$

(c) $\mathbf{x}(t_1)^T \mathbf{p}(t_1) = \int_{t_1}^{t_f} \mathbf{x}_h(t)^T \hat{\mathbf{Q}}(t) \mathbf{x}_g(t) \mathrm{d}t$

Proof 6.11

(a) The solution to Eq. (6.49a) is $\mathbf{x}_h(t) = \mathbf{\Phi}(t,t_1)\mathbf{x}(t_1)$. The solution to Eq. (6.49c) is $\mathbf{P}(t_1) = \int_{t_1}^{t_f} \mathbf{\Phi}(t,t_1)^T \hat{\mathbf{Q}}(t)\mathbf{\Phi}(t,t_1)\mathrm{d}t$. Hence

$$\frac{1}{2}\int_{t_1}^{t_f} \mathbf{x}_h(t)^T \hat{\mathbf{Q}}(t)\mathbf{x}_h(t)\mathrm{d}t = \frac{1}{2}\int_{t_1}^{t_f} \mathbf{x}(t_1)^T \mathbf{\Phi}(t,t_1)^T \hat{\mathbf{Q}}(t)\mathbf{\Phi}(t,t_1)\mathbf{x}(t_1)\mathrm{d}t$$

$$= \frac{1}{2}\mathbf{x}(t_1)^T \left(\int_{t_1}^{t_f} \mathbf{\Phi}(t,t_1)^T \hat{\mathbf{Q}}(t)\mathbf{\Phi}(t,t_1)\mathrm{d}t \right) \mathbf{x}(t_1) = \frac{1}{2}\mathbf{x}(t_1)^T \mathbf{P}(t_1)\mathbf{x}(t_1)$$

(b) Let $I_{eq} \triangleq \int_{t_1}^{t_f} \mathbf{p}(t)^T \mathbf{g}(t)\mathrm{d}t$ and $I \triangleq \frac{1}{2}\int_{t_1}^{t_f} \mathbf{x}_g(t)^T \hat{\mathbf{Q}}(t)\mathbf{x}_g(t)\mathrm{d}t$. It will be shown that $I = I_{eq}$. The solution to Eq. (6.49d) is $\mathbf{p}(t) = \int_{t}^{t_f} \mathbf{\Phi}(\tau,t)^T \mathbf{P}(\tau)\mathbf{g}(\tau)\mathrm{d}\tau$. Substituting it into I_{eq} and after that substitution of $\mathbf{P}(\tau) = \int_{\tau}^{t_f} \mathbf{\Phi}(\sigma,\tau)^T \hat{\mathbf{Q}}(\sigma)\mathbf{\Phi}(\sigma,\tau)\mathrm{d}\sigma$ yields:

$$I_{eq} = \int_{t_1}^{t_f} \left(\int_{t}^{t_f} \mathbf{\Phi}(\tau,t)^T \left(\int_{\tau}^{t_f} \mathbf{\Phi}(\sigma,\tau)^T \hat{\mathbf{Q}}(\sigma)\mathbf{\Phi}(\sigma,\tau)\mathrm{d}\sigma \right) \mathbf{g}(\tau)\mathrm{d}\tau \right)^T \mathbf{g}(t)\mathrm{d}t$$

$$= \int_{t_1}^{t_f}\mathrm{d}t \int_{t}^{t_f}\mathrm{d}\tau \int_{\tau}^{t_f}\mathrm{d}\sigma \mathbf{g}(\tau)^T \mathbf{\Phi}(\sigma,\tau)^T \hat{\mathbf{Q}}(\sigma)\mathbf{\Phi}(\sigma,\tau)\mathbf{\Phi}(\tau,t)\mathbf{g}(t)$$

However, by $\mathbf{\Phi}$'s properties, $\mathbf{\Phi}(\sigma,\tau)\mathbf{\Phi}(\tau,t) = \mathbf{\Phi}(\sigma,t)$. Hence

$$I_{eq} = \int_{t_1}^{t_f}\mathrm{d}t \int_{t}^{t_f}\mathrm{d}\tau \int_{\tau}^{t_f}\mathrm{d}\sigma \mathbf{g}(\tau)^T \mathbf{\Phi}(\sigma,\tau)^T \hat{\mathbf{Q}}(\sigma)\mathbf{\Phi}(\sigma,t)\mathbf{g}(t)$$

$$= \int_{t_1}^{t_f}\mathrm{d}t \int_{t}^{t_f}\mathrm{d}\tau \int_{\tau}^{t_f}\mathrm{d}\sigma f(\sigma,\tau,t)$$

where $f(\sigma,\tau,t) \triangleq \mathbf{g}(\tau)^T \mathbf{\Phi}(\sigma,\tau)^T \hat{\mathbf{Q}}(\sigma)\mathbf{\Phi}(\sigma,t)\mathbf{g}(t)$. Note that $f(\sigma,\tau,t) = f(\sigma,t,\tau)$.

The solution to Eq. (6.49b) is $\mathbf{x}_g(t) = \int_{t_1}^{t} \boldsymbol{\Phi}(t,\tau)\mathbf{g}(\tau)\mathrm{d}\tau$. Substituting into I leads to:

$$I = \frac{1}{2}\int_{t_1}^{t_f} \left(\int_{t_1}^{t} \boldsymbol{\Phi}(t,\tau)\mathbf{g}(\tau)\mathrm{d}\tau\right)^T \hat{\mathbf{Q}}(t)\left(\int_{t_1}^{t}\boldsymbol{\Phi}(t,\sigma)\mathbf{g}(\sigma)\mathrm{d}\sigma\right)\mathrm{d}t$$

$$= \frac{1}{2}\int_{t_1}^{t_f}\mathrm{d}t\int_{t_1}^{t}\mathrm{d}\tau\int_{t_1}^{t}\mathrm{d}\sigma\mathbf{g}(\tau)^T\boldsymbol{\Phi}(t,\tau)^T\hat{\mathbf{Q}}(t)\boldsymbol{\Phi}(t,\sigma)\mathbf{g}(\sigma)$$

As the name of the integration variable is unimportant, σ and t can be switched. It leads to

$$I = \frac{1}{2}\int_{t_1}^{t_f}\mathrm{d}\sigma\int_{t_1}^{\sigma}\mathrm{d}\tau\int_{t_1}^{\sigma}\mathrm{d}t\mathbf{g}(\tau)^T\boldsymbol{\Phi}(\sigma,\tau)^T\hat{\mathbf{Q}}(\sigma)\boldsymbol{\Phi}(\sigma,t)\mathbf{g}(t)$$

$$= \frac{1}{2}\int_{t_1}^{t_f}\mathrm{d}\sigma\int_{t_1}^{\sigma}\mathrm{d}\tau\int_{t_1}^{\sigma}\mathrm{d}t f(\sigma,\tau,t)$$

After changing the integration's order and updating its limits accordingly:

$$I = \frac{1}{2}\int_{t_1}^{t_f}\mathrm{d}t\int_{t_1}^{t_f}\mathrm{d}\tau\int_{\max(t,\tau)}^{t_f}\mathrm{d}\sigma f(\sigma,\tau,t)$$

which can be divided into two integrals by splitting the integration's limits:

$$I = \frac{1}{2}\int_{t_1}^{t_f}\mathrm{d}t\left(\int_{t_1}^{t}\mathrm{d}\tau + \int_{t}^{t_f}\mathrm{d}\tau\right)\int_{\max(t,\tau)}^{t_f}\mathrm{d}\sigma f(\sigma,\tau,t)$$

$$= \frac{1}{2}\int_{t_1}^{t_f}\mathrm{d}t\int_{t_1}^{t}\mathrm{d}\tau\int_{\max(t,\tau)}^{t_f}\mathrm{d}\sigma f(\sigma,\tau,t) + \frac{1}{2}\int_{t_1}^{t_f}\mathrm{d}t\int_{t}^{t_f}\mathrm{d}\tau\int_{\max(t,\tau)}^{t_f}\mathrm{d}\sigma f(\sigma,\tau,t)$$

For the integral on the left $\tau \in [t_1,t]$, hence the lower limit of the inner integral is $\max_{\tau\in[t_1,t]}(t,\tau) = t$. For the integral on the right $\tau \in [t,t_f]$, hence the lower limit of the inner integral is $\max_{\tau\in[t,t_f]}(t,\tau) = \tau$. Therefore

$$I = \frac{1}{2}\int_{t_1}^{t_f}\mathrm{d}t\int_{t_1}^{t}\mathrm{d}\tau\int_{t}^{t_f}\mathrm{d}\sigma f(\sigma,\tau,t) + \frac{1}{2}\int_{t_1}^{t_f}\mathrm{d}t\int_{t}^{t_f}\mathrm{d}\tau\int_{\tau}^{t_f}\mathrm{d}\sigma f(\sigma,\tau,t)$$

$$= \frac{1}{2}\int_{t_1}^{t_f}\mathrm{d}t\int_{t_1}^{t}\mathrm{d}\tau\int_{t}^{t_f}\mathrm{d}\sigma f(\sigma,\tau,t) + \frac{1}{2}I_{eq}$$

As $f(\sigma,\tau,t) = f(\sigma,t,\tau)$:

$$I = \frac{1}{2}\int_{t_1}^{t_f} \mathrm{d}t \int_{t_1}^{t} \mathrm{d}\tau \int_{t}^{t_f} \mathrm{d}\sigma f(\sigma,t,\tau) + \frac{1}{2}I_{eq}$$

As the name of the variable of integration is unimportant, t and τ can be switched:

$$I = \frac{1}{2}\int_{t_1}^{t_f} \mathrm{d}\tau \int_{t_1}^{\tau} \mathrm{d}t \int_{\tau}^{t_f} \mathrm{d}\sigma f(\sigma,\tau,t) + \frac{1}{2}I_{eq}$$

Finally, by changing the order of integration and updating the integration limits:

$$I = \frac{1}{2}\int_{t_1}^{t_f} \mathrm{d}t \int_{t}^{t_f} \mathrm{d}\tau \int_{\tau}^{t_f} \mathrm{d}\sigma f(\sigma,\tau,t) + \frac{1}{2}I_{eq} = \frac{1}{2}I_{eq} + \frac{1}{2}I_{eq} = I_{eq}$$

(c) By the explicit forms of $\mathbf{x}_h$ and $\mathbf{x}_g$, given in the proofs for (a) and (b):

$$\int_{t_1}^{t_f} \mathbf{x}_h(t)^T \hat{\mathbf{Q}}(t)\mathbf{x}_g(t)\mathrm{d}t = \int_{t_1}^{t_f} \mathbf{x}(t_1)^T \mathbf{\Phi}(t,t_1)^T \hat{\mathbf{Q}}(t) \left(\int_{t_1}^{t} \mathbf{\Phi}(t,\tau)\mathbf{g}(\tau)\mathrm{d}\tau \right) \mathrm{d}t$$

$$= \mathbf{x}(t_1)^T \int_{t_1}^{t_f} \mathrm{d}t \int_{t_1}^{t} \mathrm{d}\tau \mathbf{\Phi}(t,t_1)^T \hat{\mathbf{Q}}(t)\mathbf{\Phi}(t,\tau)\mathbf{g}(\tau) \quad (6.50)$$

On the other hand, by the explicit form of $\mathbf{P}$:

$$\mathbf{x}(t_1)^T \mathbf{p}(t_1) = \mathbf{x}(t_1)^T \int_{t_1}^{t_f} \mathbf{\Phi}(\tau,t_1)^T \mathbf{P}(\tau)\mathbf{g}(\tau)\mathrm{d}\tau$$

$$= \mathbf{x}(t_1)^T \int_{t_1}^{t_f} \mathbf{\Phi}(\tau,t_1)^T \left(\int_{\tau}^{t_f} \mathbf{\Phi}(\sigma,\tau)^T \hat{\mathbf{Q}}(\sigma)\mathbf{\Phi}(\sigma,\tau)\mathrm{d}\sigma \right) \mathbf{g}(\tau)\mathrm{d}\tau$$

$$= \mathbf{x}(t_1)^T \int_{t_1}^{t_f} \mathrm{d}\tau \int_{\tau}^{t_f} \mathrm{d}\sigma \mathbf{\Phi}(\tau,t_1)^T \mathbf{\Phi}(\sigma,\tau)^T \hat{\mathbf{Q}}(\sigma)\mathbf{\Phi}(\sigma,\tau)\mathbf{g}(\tau)$$

$$= \mathbf{x}(t_1)^T \int_{t_1}^{t_f} \mathrm{d}\tau \int_{\tau}^{t_f} \mathrm{d}\sigma \mathbf{\Phi}(\sigma,t_1)^T \hat{\mathbf{Q}}(\sigma)\mathbf{\Phi}(\sigma,\tau)\mathbf{g}(\tau)$$

where in the last line the property $\boldsymbol{\Phi}(\sigma,\tau)\boldsymbol{\Phi}(\tau,t_1) = \boldsymbol{\Phi}(\sigma,t_1)$ was used. By writing t instead of σ:

$$\mathbf{x}(t_1)^T\mathbf{p}(t_1) = \mathbf{x}(t_1)^T \int\limits_{t_1}^{t_f} d\tau \int\limits_{\tau}^{t_f} dt\boldsymbol{\Phi}(t,t_1)^T\hat{\mathbf{Q}}(t)\boldsymbol{\Phi}(t,\tau)\mathbf{g}(\tau)$$

Changing the integration's order and updating its limits leads to:

$$\mathbf{x}(t_1)^T\mathbf{p}(t_1) = \mathbf{x}(t_1)^T \int\limits_{t_1}^{t_f} dt \int\limits_{t_1}^{t} d\tau\boldsymbol{\Phi}(t,t_1)^T\hat{\mathbf{Q}}(t)\boldsymbol{\Phi}(t,\tau)\mathbf{g}(\tau)$$

which is identical to Eq. (6.50).

Lemma 6.11

Let $\mathbf{x}_h$ und $\mathbf{x}_g$ be defined by Eqs. (6.49a) and (6.49b). Let $\mathbf{P}_x$, $\mathbf{P}_u$ and $\mathbf{P}_{xu}$ be solutions of Eq. (6.49c) for $\hat{\mathbf{Q}}$s, defined by

$$\hat{\mathbf{Q}}(t) = \mathbf{Q}$$
$$\hat{\mathbf{Q}}(t) = \mathbf{C}^T\,\mathbf{diag}((u_i(t)^2 r_i)_{i=1}^{n_u})\mathbf{C}$$
$$\hat{\mathbf{Q}}(t) = -\mathbf{N}^T\,\mathbf{diag}(\mathbf{u}(t))\mathbf{C} - \mathbf{C}^T\,\mathbf{diag}(\mathbf{u}(t))\mathbf{N}$$

respectively. Let $\mathbf{p}_x$, $\mathbf{p}_u$ and $\mathbf{p}_{xu}$ be solutions of Eq. (6.49d) for $\mathbf{P}_x$, $\mathbf{P}_u$ and $\mathbf{P}_{xu}$, respectively. Then:

(a) $\frac{1}{2}\mathbf{x}(t_1)^T\mathbf{P}_x(t_1)\mathbf{x}(t_1) = \frac{1}{2}\int\limits_{t_1}^{t_f}\mathbf{x}_h(t)^T\mathbf{Q}\mathbf{x}_h(t)\mathrm{d}t$

(b) $\frac{1}{2}\mathbf{x}(t_1)^T\mathbf{P}_u(t_1)\mathbf{x}(t_1) = \frac{1}{2}\int\limits_{t_1}^{t_f}\mathbf{x}_h(t)^T\mathbf{C}^T\,\mathbf{diag}((u_i(t)^2 r_i)_{i=1}^{n_u})\mathbf{C}\mathbf{x}_h(t)\mathrm{d}t$

(c) $\frac{1}{2}\mathbf{x}(t_1)^T\mathbf{P}_{xu}(t_1)\mathbf{x}(t_1) = -\int\limits_{t_1}^{t_f}\mathbf{x}_h(t)^T\mathbf{N}^T\,\mathbf{diag}(\mathbf{u}(t))\mathbf{C}\mathbf{x}_h(t)\mathrm{d}t$

(d) $\int\limits_{t_1}^{t_f}\mathbf{p}_x(t)^T\mathbf{g}(t)\mathrm{d}t = \frac{1}{2}\int\limits_{t_1}^{t_f}\mathbf{x}_g(t)^T\mathbf{Q}\mathbf{x}_g(t)\mathrm{d}t$

(e) $\int\limits_{t_1}^{t_f}\mathbf{p}_u(t)^T\mathbf{g}(t)\mathrm{d}t = \frac{1}{2}\int\limits_{t_1}^{t_f}\mathbf{x}_g(t)^T\mathbf{C}^T\,\mathbf{diag}((u_i(t)^2 r_i)_{i=1}^{n_u})\mathbf{C}\mathbf{x}_g(t)\mathrm{d}t$

(f) $\int\limits_{t_1}^{t_f}\mathbf{p}_{xu}(t)^T\mathbf{g}(t)\mathrm{d}t = -\int\limits_{t_1}^{t_f}\mathbf{x}_g(t)^T\mathbf{N}^T\,\mathbf{diag}(\mathbf{u}(t))\mathbf{C}\mathbf{x}_g(t)\mathrm{d}t$

(g) $\mathbf{x}(t_1)^T\mathbf{p}_x(t_1) = \int\limits_{t_1}^{t_f}\mathbf{x}_h(t)^T\mathbf{Q}\mathbf{x}_g(t)\mathrm{d}t$

(h) $\quad \mathbf{x}(t_1)^T \mathbf{p}_u(t_1) = \int_{t_1}^{t_f} \mathbf{x}_h(t)^T \mathbf{C}^T \mathbf{diag}((u_i(t)^2 r_i)_{i=1}^{n_u}) \mathbf{C} \mathbf{x}_g(t) \mathrm{d}t$

(i) $\quad \mathbf{x}(t_1)^T \mathbf{p}_{xu}(t_1) = -2 \int_{t_1}^{t_f} \mathbf{x}_h(t)^T \mathbf{N}^T \mathbf{diag}(\mathbf{u}(t)) \mathbf{C} \mathbf{x}_g(t) \mathrm{d}t$

Proof 6.12 (a), (d) and (g) are proved by Lemma 6.10 when $\hat{\mathbf{Q}}(t) = \mathbf{Q}$. (b), (e) and (h) are proved by Lemma 6.10 when $\hat{\mathbf{Q}}(t) = \mathbf{C}^T \mathbf{diag}((u_i(t)^2 r_i)_{i=1}^{n_u}) \mathbf{C}$. (c), (f) and (i) are proved by recalling that for $\boldsymbol{\xi} \in \mathbb{R}^n$

$$-\boldsymbol{\xi}^T \mathbf{N}^T \mathbf{diag}(\mathbf{u}(t)) \mathbf{C} \boldsymbol{\xi} = \frac{1}{2} \boldsymbol{\xi}^T (-\mathbf{N}^T \mathbf{diag}(\mathbf{u}(t)) \mathbf{C} - \mathbf{C}^T \mathbf{diag}(\mathbf{u}(t)) \mathbf{N}) \boldsymbol{\xi}$$

and then using Lemma 6.10 with $\hat{\mathbf{Q}}(t) = (-\mathbf{N}^T \mathbf{diag}(\mathbf{u}(t)) \mathbf{C} - \mathbf{C}^T \mathbf{diag}(\mathbf{u}(t)) \mathbf{N})$.

Example 6.3.1: Single input CBBR

Consider a free vibrating mass-spring system, shown in Fig. 6.9. It is composed of three bodies, connected by three identical springs. The mass of each body is 10^5 [kg] and the stiffness of each spring is $3.6 \cdot 10^6$ [N/m]. The control force is applied to the first body by a single semi-active actuator. The state space equation is:

$$\frac{d}{dt} \begin{bmatrix} z_1(t) \\ z_2(t) \\ z_3(t) \\ \dot{z}_1(t) \\ \dot{z}_2(t) \\ \dot{z}_3(t) \end{bmatrix} = \begin{bmatrix} 0 & 0 & 0 & 1 & 0 & 0 \\ 0 & 0 & 0 & 0 & 1 & 0 \\ 0 & 0 & 0 & 0 & 0 & 1 \\ -72 & 36 & 0 & 0 & 0 & 0 \\ 36 & -72 & 36 & 0 & 0 & 0 \\ 0 & 36 & -36 & 0 & 0 & 0 \end{bmatrix} \begin{bmatrix} z_1(t) \\ z_2(t) \\ z_3(t) \\ \dot{z}_1(t) \\ \dot{z}_2(t) \\ \dot{z}_3(t) \end{bmatrix} + \begin{bmatrix} 0 \\ 0 \\ 0 \\ 10^{-5} \\ 0 \\ 0 \end{bmatrix} w(t) \quad (6.51)$$

where $z_2(0) = 0.6$ and $z_3(0) = -0.6$. Three cases are studied:

Case 1: An uncontrolled plant.

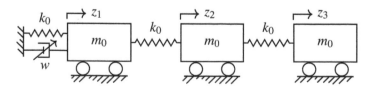

Figure 6.9: The plant's model [44].

Case 2: A CBBR controlled plant without bound on the control force magnitude.

Case 3: A CBBR controlled plant with $w_{max} = 4 \times 10^5$ [N].

As $z_d = z_1$, case 3's constraints are (C1) $w(t)\dot{z}_1(t) \leq 0$; (C2) $\dot{z}_1(t) = 0 \rightarrow w(t) = 0$; and (C3) $|w(t)| \leq 4 \times 10^5$, for all $t \in [0,8]$. Case 2's constraints are the same, only that (C3) is omitted. The performance index is

$$J(\mathbf{x},w) = \frac{1}{2} \int_0^8 \left(z_1(t)^2 + (z_2(t) - z_1(t))^2 + (z_3(t) - z_2(t))^2 + 10^{-13}w(t)^2 \right) dt$$

(6.52)

19 iterations were performed in the CBBR design. Performance index values per iteration are illustrated in Fig. 6.10. As it is evident from this figure, case 2 converged after 5 iterations and case 3 converged after 2 iterations. For case 3 the major improvement was achieved by the first iteration. In both cases, a monotonic reduction in J_i values is evident. Displacements in z_3, for cases 1–3, are presented in Fig. 6.11. The most significant improvement is for case 2. This is not surprising, as for this case the control force is unbounded. The control force is depicted in Fig. 6.12. During the first part of the control duration, the signal form of case 3 is very similar to that of signals derived by the *bang-bang control* approach. This is because the response intensity in that part generated large control force demand, thereby causing force saturation. Figure 6.13 illustrates the correspondence between the actuator's control force and velocity in cases 2 and 3. Each pair $(\dot{z}_d(t), w(t))$ is represented by a point on the graph. For both cases there are no points in the 1st and 3rd quarter of the $\dot{z} - w$ plane. This infers that the control power is always negative. It can also be seen that the control force magnitude for case 3 does not exceed $\pm w_{max}$. It follows that constraints C1 and C3, from Definition 6.2, are satisfied. Figure 6.14 describes the hysteresis plots for cases 2 and 3, i.e., the correspondence between the actuator's control force and elongation. It should be noted that for both cases significant control forces are generated at peak displacements, which imply that the control forces in these cases are in-phase with the displacement. The control trajectories u for cases 2 and 3 are given in Fig. 6.15. This figure provides a clear illustration of the trajectories' discontinuous form. Figures 6.16 and 6.17 illustrate the form of elements $(\mathbf{P})_{11}$ and $(\mathbf{P})_{12}$ from the matrix function $\mathbf{P}$.

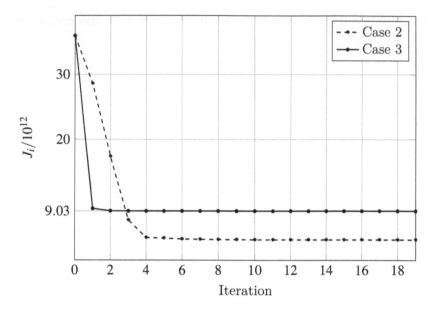

Figure 6.10: Performance index values per iteration.

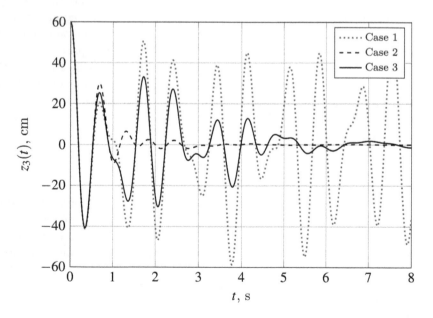

Figure 6.11: The displacement's time history in the 3rd DOF.

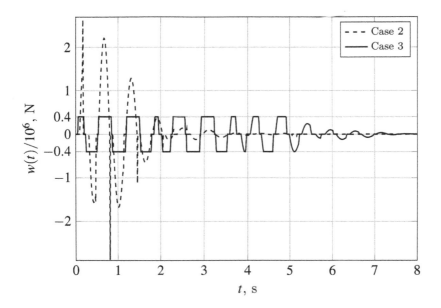

Figure 6.12: The control force's time history.

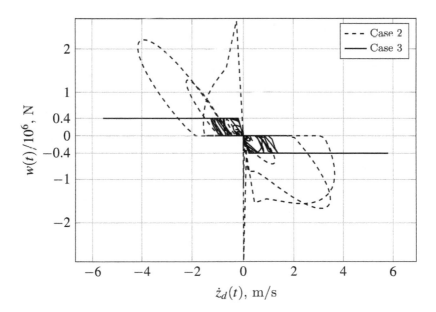

Figure 6.13: Control force $w(t)$ vs. actuator velocity $\dot{z}_d(t)$.

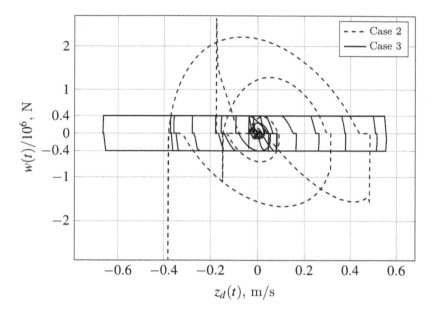

Figure 6.14: Damper hysteresis.

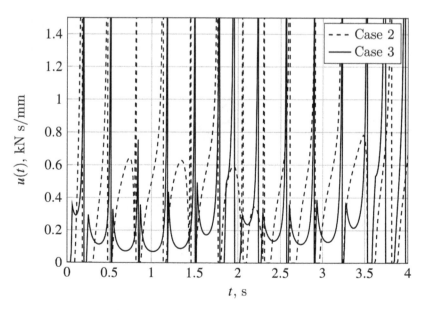

Figure 6.15: Control trajectories.

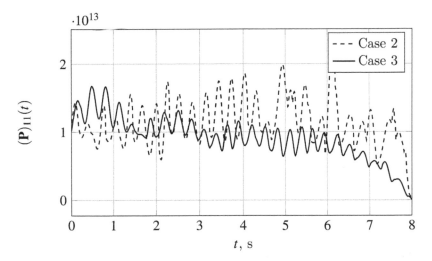

Figure 6.16: Lyapunov solution—element $(\mathbf{P})_{11}(t)$.

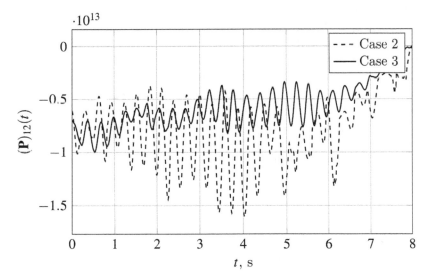

Figure 6.17: Lyapunov solution—element $(\mathbf{P})_{12}(t)$.

Example 6.3.2: Multi-input CBBR

This example illustrates a CBBR solution at a multi-control input case. It follows a benchmark problem for structures that are subjected to wind excitation, proposed by Yang et al.[105]. The plant is a RC structure that was destined to be an office tower in Melbourne, Australia. However, it was not actually built due to an economic recession. The building consists of 76-stories and its height is 306 [m]. The dynamic scheme and a typical floor cross section are given in the original paper [105]. The structural system consists of two main parts: a concrete core and a concrete frame. The core was designed to resist the majority of wind loads, whereas the frame—the gravitational loads and part of the wind loads. The floor cross section is square with chamfers in two corners. The total mass of the building, including heavy machinery in six plant rooms, is 153,000 [ton]. The RC perimeter frame consists of 24 columns at each level, connected to a 900 [mm] deep and 400 [mm] wide spandrel beam. The lightweight floor construction uses steel beams with a metal deck and a 120 [mm] slab. Column sizes, core wall thickness and floor mass vary along the height. The concrete's compressive strength is 60 [MPa] and its elastic modulus is 40 [GPa]. The building is modeled as a vertical cantilever beam. First, a finite element model was constructed by considering the portion of the building between two adjacent floors as a classical beam element of uniform thickness, leading to 76 translational and 76 rotational DOFs. Next, the rotational DOFs were removed by Guyan reduction, resulting in a model with 76 DOFs that represents the lateral displacements at each floor. The damping matrix was calculated by Rayleigh's approach, assuming 1% natural damping ratios for the first five modes. This model, having 76×76 mass, damping, and stiffness matrices, is referred to as the *76 DOFs Model*. A detailed description of the structure and its model can be found in the literature [105].

The wind force data, acting on the benchmark building, was determined by wind tunnel tests that were conducted on a 1:400 scaled model [84]. The wind pattern and the wind tunnel were set to simulate natural wind over a suburban terrain. It was observed that the building response due to across-wind loads was much more intense than that caused by the along-wind ones. Therefore only the across-wind loads were considered in the benchmark problem [105]. Additionally, in order to reduce the computational burden of the benchmark model response, only the first 900 seconds of the recorded data were used. It should be noted that the mean wind force had been removed from the wind data as it produces just a static deflection. The wind loads' sample time is 0.133 [s]. A detailed description of the wind-tunnel test and the scaled model can be found in the relevant paper [84].

Model Reduction.

The control design was carried out by considering a reduced order model. To this end, the model reduction approach, that was suggested in [29], was used. It formulates a lower order plant with the same dominant mode shapes and frequencies as the original system.

Yet, the model reduction in this example is different than that described in [105]. First, in [105] the number of wind input signals was reduced, whereas here the number of these inputs is maintained. Their influence on the reduced model is brought into consideration by appropriate modification of the wind input matrices. Second, here the model is reduced to $n = 6$ rather than $n = 46$ in [105]. Such system order reduces **P**'s stiffness, on the one hand, and provides a satisfying description of the vibration, on the other hand. Hence, there are some small differences in the response of the uncontrolled structure, compared to the original paper.

The reduced model takes into account the first 3 DOF modes (which are the first 6 state modes) and DOFs 1, 50 and 76 from the 76 DOFs model. It is subjected to 76 known wind force inputs and control force inputs from 3 semi-active dampers. The dampers are installed in floors 1 to 3, where inter-story drifts tend to be quite large. The state space equations are:

$$\dot{\mathbf{x}}(t) = \mathbf{A}\mathbf{x}(t) + \mathbf{B}\mathbf{w}(t) + \mathbf{E}\mathbf{w}_w(t); \quad \mathbf{x}(0) = \mathbf{0}, \forall t \in (0, 900)$$

Here $\mathbf{x} = \begin{bmatrix} z_1 & z_{50} & z_{76} & \dot{z}_1 & \dot{z}_{50} & \dot{z}_{76} \end{bmatrix}^T$ is the state trajectory of the reduced model; $\mathbf{w} = \begin{bmatrix} w_1 & w_2 & w_3 \end{bmatrix}^T$ is the control forces trajectory; $\mathbf{w}_w : \mathbb{R} \to \mathbb{R}^{76}$ is the wind forces trajectory; and $\mathbf{A} \in \mathbb{R}^{6 \times 6}$, $\mathbf{B} \in \mathbb{R}^{6 \times 3}$ and $\mathbf{E} \in \mathbb{R}^{6 \times 76}$ are corresponding matrices that are formulated by the above mentioned model reduction method. The initial conditions are zero.

The form of the excitation input $\mathbf{g} = \mathbf{E}\mathbf{w}_w$ is illustrated in Fig. 6.18. Only g_4, g_5 and g_6 are described as the others are equal to zero.

Two cases are considered:

Case 1: Uncontrolled plant.

Case 2: CBBR controlled plant.

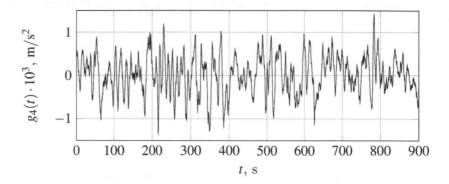

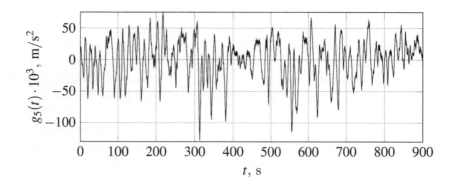

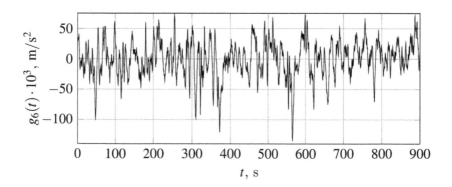

Figure 6.18: Excitations g_4, g_5 and g_6.

Let $\mathbf{z} = (z_i)_{i=1}^{76}$ be the DOFs trajectory of the reduced model. That is, z_1, z_{50} and z_{76} are taken from $\mathbf{x}$ and the rest of the DOFs are obtained as explained in [29]. The matrix $\mathbf{C} \in \mathbb{R}^{3 \times 6}$ is formulated such that $\mathbf{c}_1 \mathbf{x} = \dot{z}_1$, $\mathbf{c}_2 \mathbf{x} = \dot{z}_2 - \dot{z}_1$ and $\mathbf{c}_3 \mathbf{x} = \dot{z}_3 - \dot{z}_2$. The control constraints are (C1) $w_i(t)\mathbf{c}_i\mathbf{x}(t) \leq 0$; (C2) $\mathbf{c}_i\mathbf{x}(t) = 0 \rightarrow w_i(t) = 0$; and (C3) $|w(t)| \leq 50 \times 10^3$, for all $t \in [0, 900]$ and $i \in 1, 2, 3$.

The plant performance is evaluated with respect to an output $\mathbf{y} : \mathbb{R} \rightarrow \mathbb{R}^{27}$ of the form:

$$\mathbf{y}(t) = \mathbf{C}_y \mathbf{x}(t) + \mathbf{D}\mathbf{w}(t)$$

where $\mathbf{C}_y \in \mathbb{R}^{27 \times 6}$ and $\mathbf{D} \in \mathbb{R}^{27 \times 3}$ are written such that

$$\mathbf{y} \triangleq \begin{bmatrix} \mathbf{z}_\eta \\ \dot{\mathbf{z}}_\eta \\ (\ddot{\mathbf{z}} - \mathbf{\Phi}_w \mathbf{w}_w)_\eta \end{bmatrix}$$

Here $\mathbf{z}_\eta \triangleq (z_i)_{i=\eta}$, where $\eta = \{1, 30, 50, 55, 60, 65, 70, 75, 76\}$. The performance index is

$$J(\mathbf{y}, \mathbf{w}) = \int_0^{t_f} \left(\mathbf{y}(t)^T \mathbf{Q}_0 \mathbf{y}(t) + 20.15\mathbf{w}(t)^T \mathbf{w}(t) \right) dt$$

where $\mathbf{Q}_0 \triangleq \mathbf{diag}(1, 1, 1, 1, 1, 1, 1, 1, 1, 1, 1, 1, 1, 1, 10^5, 10^5, 10^5, 10^5, 1, 1, 1, 1,$ $1, 10^5, 10^5, 10^5, 10^5) \cdot 10^{10}$. Substitution of the output equation yields a performance index of the form

$$J(\mathbf{x}, \mathbf{w}) = \frac{1}{2} \int_0^{t_f} \left(\mathbf{x}(t)^T \mathbf{Q}\mathbf{x}(t) + \mathbf{w}(t)^T \mathbf{R}\mathbf{w}(t) + 2\mathbf{w}(t)^T \mathbf{N}\mathbf{x}(t) \right) dt$$

$$\mathbf{Q} = 2\mathbf{C}_y^T \mathbf{Q}_0 \mathbf{C}_y; \ \mathbf{R} = 2(\mathbf{D}^T \mathbf{Q}_0 \mathbf{D} + 20.15\mathbf{I}); \ \mathbf{N} = 2\mathbf{D}^T \mathbf{Q}_0 \mathbf{C}_y$$

7 iteration where carried out during the CBBR design. Performance index values per iteration are illustrated in Fig. 6.19. The control force trajectories for each damper are shown in Fig. 6.20. It can be seen that the force signals saturate due to constraint C3. The control trajectories for each damper are given in Fig. 6.21. Figures 6.22 and 6.23 present the form of elements $(\mathbf{P})_{11}$ and $(\mathbf{P})_{12}$ from the matrix function $\mathbf{P}$.

The control efficiency in case 2 is evaluated by twelve criteria, suggested in the original benchmark problem [105] and briefly described here in Table 6.1. A tilde sign is used to distinguish these criteria from the performance index J. A notation σ_z is used in the table to denote the

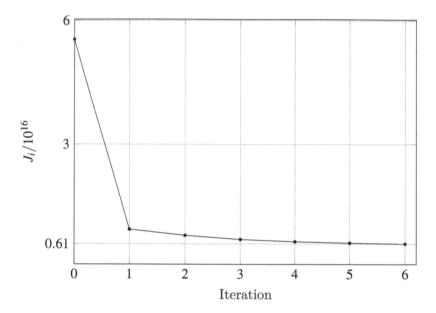

Figure 6.19: Performance index values per iteration.

root-mean-square (RMS) of a trajectory $z: [0, t_f] \to \mathbb{R}$, i.e.:

$$\sigma_z \triangleq \sqrt{\frac{1}{t_f} \int_0^{t_f} z(t)^2 \mathrm{d}t}$$

and $\|z\|_\infty$ is used to denote the infinity norm of z, i.e., $\sup_{t \in [0, t_f]} |z(t)|$. Trajectories from case 1 will be denoted in this table by a subscript 'o'.

The values of the benchmark criteria for case 2 are given in Table 6.2. It can be seen from that table that the control efficiently improves the plant performance. However, it should be emphasized that it cannot be compared to the benchmark results from [105] because, as it was explained above, the plant under consideration is not identical to the original one.

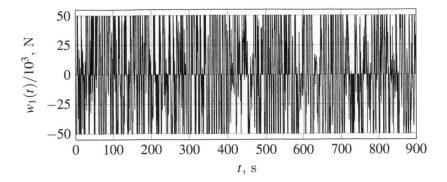

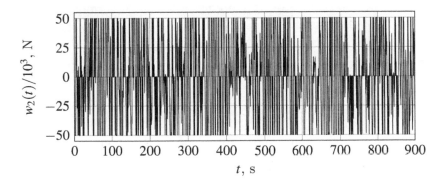

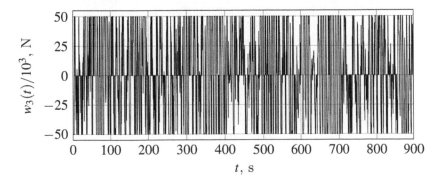

Figure 6.20: Control force trajectories w_1, w_2 and w_3.

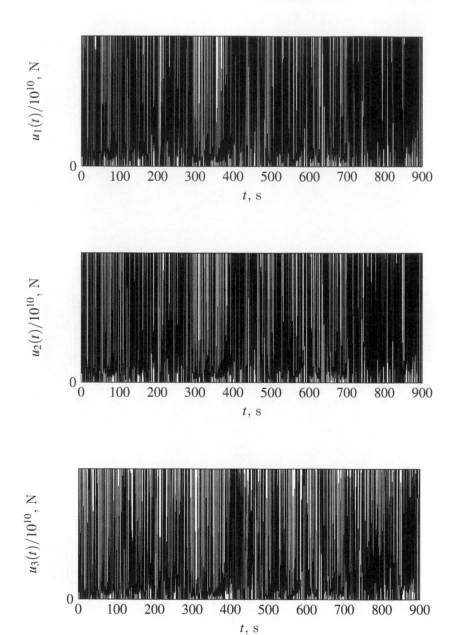

Figure 6.21: Control trajectories u_1, u_2 and u_3.

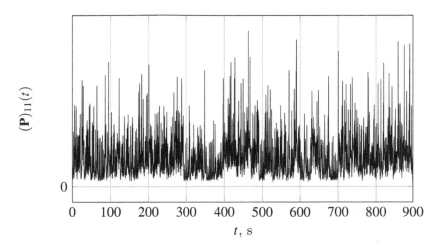

Figure 6.22: Lyapunov equation's solution—element $(\mathbf{P})_{11}(t)$.

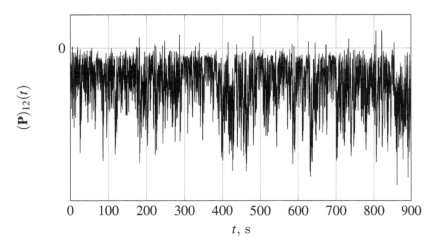

Figure 6.23: Lyapunov equation's solution—element $(\mathbf{P})_{12}(t)$.

Table 6.1: Benchmark criteria; σ_z is the RMS of z; $V_1 = \{1, 30, 50, 55, 60, 65, 70, 75\}$; $V_2 = \{50, 55, 60, 65, 70, 75\}$ and $V_3 = \{50, 55, 60, 65, 70, 75, 76\}$.

Benchmark Criterion	Meaning
$\tilde{J}_1 = \max_{i \in V_1} \left(\frac{\sigma_{\ddot{z}_i}}{\sigma_{\ddot{z}_{o75}}} \right)$	Max. floor RMS acceleration
$\tilde{J}_2 = \frac{1}{6} \sum_{i \in V_2} \frac{\sigma_{\ddot{z}_i}}{\sigma_{\ddot{z}_{oi}}}$	Average acceleration performance at top floors.
$\tilde{J}_3 = \frac{\sigma_{z_{76}}}{\sigma_{z_{o76}}}$	Reduction at top floor displacements.
$\tilde{J}_4 = \frac{1}{7} \sum_{i \in V_3} \frac{\sigma_{z_i}}{\sigma_{z_{oi}}}$	Average displacements performance in top floors.
$\tilde{J}_5 = \left(\frac{\sigma_{z_{di}}}{\sigma_{z_{o76}}} \right)^{n_u}_{i=1}$	Nondimensionalized RMS of actuators' stroke.
$\tilde{J}_6 = (\sigma_{\pi_i})^{n_u}_{i=1};$ $\pi_i(t) \triangleq w_i(t)\dot{z}_{di}(t)$	RMS of actuators' power.
$\tilde{J}_7 = \max_{i \in V_1} \left(\frac{\|\ddot{z}_i\|_\infty}{\|\ddot{z}_{o75}\|_\infty} \right)$	Max. peak floor acceleration
$\tilde{J}_8 = \frac{1}{6} \sum_{i \in V_2} \frac{\|\ddot{z}_i\|_\infty}{\|\ddot{z}_{oi}\|_\infty}$	Average peak acceleration at top floors.
$\tilde{J}_9 = \frac{\|z_{76}\|_\infty}{\|z_{o76}\|_\infty}$	Reduction in peak top floor displacements.
$\tilde{J}_{10} = \frac{1}{7} \sum_{i \in V_3} \frac{\|z_i\|_\infty}{\|z_{oi}\|_\infty}$	Average peak displacements at top floors.
$\tilde{J}_{11} = \left(\frac{\|z_{di}\|_\infty}{\|z_{o76}\|_\infty} \right)^{n_u}_{i=1}$	Nondimensionalized actuators' peak stroke.
$\tilde{J}_{12} = (\|\pi_i\|_\infty)^{n_u}_{i=1}$	Actuators' peak power.

Table 6.2: Benchmark criteria values for case 2.

Criterion	Value
$\tilde{J}_1$	0.331388
$\tilde{J}_2$	0.306785
$\tilde{J}_3$	0.457211
$\tilde{J}_4$	0.459348
$\tilde{J}_5$	$\begin{bmatrix} 0.000820356 & 0.00110924 & 0.00118986 \end{bmatrix}$
$\tilde{J}_6$	$\begin{bmatrix} 1.93017 & 2.077 & 2.76033 \end{bmatrix}$
$\tilde{J}_7$	0.591537
$\tilde{J}_8$	0.546804
$\tilde{J}_9$	0.591205
$\tilde{J}_{10}$	0.596722
$\tilde{J}_{11}$	$\begin{bmatrix} 0.00107309 & 0.00145034 & 0.00155604 \end{bmatrix}$
$\tilde{J}_{12}$	$\begin{bmatrix} 16.0067 & 16.9114 & 19.5675 \end{bmatrix}$

6.4 Constrained Bilinear Quadratic Regulator

Some semi-active dampers, e.g. variable viscous damper [78] or variable orifice damper [65], are characterized by a controllable viscous gain. In such case it might be logical to pay more attention to the viscous gain rather than to the control force itself. Hence, unlike in CBBR, here a constrained optimal control problem is formulated with emphasis on the controlled viscous gain. In the next problem statement, the control forces are not considered explicitly by the performance index, however, they can be bounded by accordingly defined constraints. It is interesting to see that the different formulation of the performance index affects the control signals' shaping. The obtained control forces are generally similar to those generated by VDs, and benefit out of the phase control forces, as in simple VDs. Such optimal control design problem will be denoted here as *constrained bilinear quadratic regulator* (CBQR). First, an optimal control problem will be defined. Afterwards, a sequence of improving functions, which corresponds to the CBQR problem, will be formulated and Krotov's method will be used for solving it.

Let a control force in a semi-active damper be described as in Eq. (6.11), i.e., the control signal is the viscous gain u_i. An admissible process, in this case, is defined as follows.

Definition 6.4   Let $\mathbf{x} : \mathbb{R} \to \mathbb{R}^n$ be a state trajectory and $\mathbf{u} : \mathbb{R} \to \mathbb{R}^{n_u}$ be a control trajectory. Let $\mathscr{U}_i(\mathbf{x})$ be a set of admissible control trajectories, where each $u \in \mathscr{U}_i(\mathbf{x})$ satisfies

$$u(t) \in [u_{i,min}(t, \mathbf{x}(t)), u_{i,max}(t, \mathbf{x}(t))] \tag{6.53}$$

for all $t \in [0, t_f]$. $u_{i,min}$ and $u_{i,max}$ are mappings $\mathbb{R} \times \mathbb{R}^n \to [0, \infty)$, which satisfy $0 \leq u_{i,min}(t, \xi) \leq u_{i,max}(t, \xi)$ for all $t \in [0, t_f]$ and $\xi \in \mathbb{R}^n$.

The pair $(\mathbf{x}, \mathbf{u})$ is an admissible process if it satisfies the bilinear state equation

$$\dot{\mathbf{x}}(t) = \left(\mathbf{A} - \sum_{i=1}^{n_u} \mathbf{b}_i u_i(t) \mathbf{c}_i \right) \mathbf{x}(t) + \mathbf{g}(t); \quad \mathbf{x}(0), \forall t \in (0, t_f) \tag{6.54}$$

and $u_i \in \mathscr{U}_i(\mathbf{x})$ for all $i = 1, \dots, n_u$. Here $\mathbf{A} \in \mathbb{R}^{n \times n}$; $\mathbf{g} : \mathbb{R} \to \mathbb{R}^n$ is a trajectory of known bounded external excitations and $\mathbf{b}_i, \mathbf{c}_i^T \in \mathbb{R}^n$.

Remarks.

■ $u_{i,min}$ and $u_{i,max}$ represent bounds on the viscous gain. For dampers with linear viscous damping these bounds are some positive constants, i.e., $u_{i,min} = c_{di,min} \geq 0$ and $u_{i,max} = c_{di,max} \geq c_{di,min}$.

As it was explained at the beginning of Section 6.3, in many cases there are bounds on the damper's force magnitude such that

$$w_{i,max} \geq |w_i(t, \mathbf{x}(t))| = |u(t) \mathbf{c}_i \mathbf{x}(t)| \tag{6.55}$$

for some $w_{i,max} > 0$. In this case the viscous gain upper bound can be set as

$$u_{i,max}(t, \mathbf{x}(t)) = \begin{cases} c_{di,max} & , \mathbf{c}_i \mathbf{x}(t) = 0 \\ \min \left(c_{di,max}, \frac{w_{i,max}}{|\mathbf{c}_i \mathbf{x}(t)|} \right) & , \mathbf{c}_i \mathbf{x}(t) \neq 0 \end{cases} \tag{6.56}$$

It should be noted that when $\mathbf{c}_i \mathbf{x}(t) = 0$, $u_i(t)$ has no practical importance. This happens by virtue of C2, that is $w_i(t, \mathbf{x}(t)) = 0$ whenever

$\mathbf{c}_i\mathbf{x}(t) = 0$. It means that the viscous gain does not contribute to the process at these time instances and its value can be set arbitrarily.

Another example is controlled viscous dampers whose viscous gain law is unnecessarily linear, as in Eq. (2.9). The control force can be written as

$$w(t, \dot{z}_d(t)) = -c_d(t)|\dot{z}_d(t)|^\alpha \operatorname{sign}(\dot{z}_d(t)) \tag{6.57}$$

$$= -c_d(t)|\dot{z}_d(t)|^\alpha \frac{\dot{z}_d(t)}{|\dot{z}_d(t)|} \tag{6.58}$$

$$= -c_d(t)|\dot{z}_d(t)|^{\alpha-1}\dot{z}_d(t) \tag{6.59}$$

Hence it is possible to define $u_i(t,\mathbf{x}(t)) \triangleq c_{di}(t)|\mathbf{c}_i\mathbf{x}(t)|^{\alpha-1}$. It implies that

$$u_{i,min}(t,\mathbf{x}(t)) = c_{di,min}(t)|\mathbf{c}_i\mathbf{x}(t)|^{\alpha-1}$$
$$u_{i,max}(t,\mathbf{x}(t)) = c_{di,max}|\mathbf{c}_i\mathbf{x}(t)|^{\alpha-1}$$

However, it can be seen that such $u_{i,min}$ and $u_{i,max}$ allow for u_i to become unbounded whenever $\alpha < 1$ and $\mathbf{c}_i\mathbf{x}(t) \to 0$.

■ It is easy to see that the control forces satisfy constraints C1 and C2. C3 is satisfied by setting $w_{l,min}(t,\mathbf{x}(t)) = u_{i,min}(t,\mathbf{x}(t))|\mathbf{c}_i\mathbf{x}(t)|$.

The CBQR problem is defined as follows.

Problem 6.3 (CBQR) *The CBQR control problem is a search for an optimal and admissible process, $(\mathbf{x}^*, \mathbf{u}^*)$, that minimizes the quadratic performance index*

$$J(\mathbf{x}, \mathbf{u}) = \frac{1}{2}\int_0^{t_f} \left(\mathbf{x}(t)^T \mathbf{Q}\mathbf{x}(t) + \sum_{i=1}^{n_u} u_i(t)^2 r_i \right) dt \tag{6.60}$$

where $0 \le \mathbf{Q} \in \mathbb{R}^{n \times n}$ and $r_i > 0$ for $i = 1, \ldots, n_u$.

This problem will be solved hereinafter by Krotov's method. With this aim, a class of improving functions that suits the CBQR problem will be formulated in the next lemmas. The lemmas and their proofs are quite similar to those of Lemmas 6.4 and 6.5.

Lemma 6.12

Let

$$q(t,\xi) = \frac{1}{2}\xi^T \mathbf{P}(t)\xi + \mathbf{p}(t)^T \xi; \quad \mathbf{P}(t_f) = \mathbf{0}; \mathbf{p}(t_f) = \mathbf{0}$$

where $\xi \in \mathbb{R}^n$, $\mathbf{P} : \mathbb{R} \to \mathbb{R}^{n \times n}$ *is a continuous, piecewise smooth and symmetric, matrix function and* $\mathbf{p} : \mathbb{R} \to \mathbb{R}^n$ *is a continuous and piecewise smooth, vector function.*

Let $v_i(t,\xi) \triangleq \frac{\mathbf{b}_i^T (\mathbf{P}(t)\xi + \mathbf{p}(t))\mathbf{c}_i \xi}{r_i}$. *The vector of control laws,* $(\hat{u}_i)_{i=1}^{n_u}$, *that minimizes* $s(t, \mathbf{x}(t), \mathbf{u}(t))$, *is given by*

$$\hat{u}_i(t, \mathbf{x}(t)) = \arg \min_{\mathbf{v} \in \mathscr{U}_i(t,\mathbf{x})} s(t, \mathbf{x}(t), \mathbf{v})$$

$$= \begin{cases} u_{i,min}(t, \mathbf{x}(t)) & , v_i(t, \mathbf{x}(t)) \leq u_{i,min}(t, \mathbf{x}(t)) \\ u_{i,max}(t, \mathbf{x}(t)) & , v_i(t, \mathbf{x}(t)) \geq u_{i,max}(t, \mathbf{x}(t)) \\ v_i(t, \mathbf{x}(t)) & , otherwise \end{cases} \tag{6.61}$$

Proof 6.13 The partial derivatives of q are

$$q_t(t,\xi) = \frac{1}{2}\xi^T \dot{\mathbf{P}}(t)\xi + \dot{\mathbf{p}}(t)^T \xi \tag{6.62}$$

$$q_{\mathbf{x}}(t,\xi) = \xi^T \mathbf{P}(t) + \mathbf{p}(t)^T \tag{6.63}$$

Let $\mathbf{v} \in \mathbb{R}^{n_u}$. Substituting it into Eqs. (3.18) and (3.19) yields:

$$s_f(\mathbf{x}(t_f)) = 0 \quad \forall \mathbf{x}(t_f) \in \mathscr{X}(t_f) \tag{6.64}$$

$$\begin{aligned}
s(t, \mathbf{x}(t), \mathbf{v}) &= q_t(t, \mathbf{x}(t)) + q_{\mathbf{x}}(t, \mathbf{x}(t)) \mathbf{f}(t, \mathbf{x}(t), \mathbf{v}) \\
&\quad + \frac{1}{2} \left(\mathbf{x}(t)^T \mathbf{Q} \mathbf{x}(t) + \sum_{i=1}^{n_u} v_i^2 r_i \right)
\end{aligned} \tag{6.65}$$

$$\begin{aligned}
&= \frac{1}{2} \mathbf{x}(t)^T \dot{\mathbf{P}}(t) \mathbf{x}(t) + \dot{\mathbf{p}}(t)^T \mathbf{x}(t) + \mathbf{x}(t)^T \mathbf{P}(t) \mathbf{f}(t, \mathbf{x}(t), \mathbf{v}) \\
&\quad + \mathbf{p}(t)^T \mathbf{f}(t, \mathbf{x}(t), \mathbf{v}) + \frac{1}{2} \mathbf{x}(t)^T \mathbf{Q} \mathbf{x}(t) + \frac{1}{2} \sum_{i=1}^{n_u} v_i^2 r_i
\end{aligned} \tag{6.66}$$

$$\begin{aligned}
&= \frac{1}{2} \mathbf{x}(t)^T \dot{\mathbf{P}}(t) \mathbf{x}(t) + \dot{\mathbf{p}}(t)^T \mathbf{x}(t) \\
&\quad + \mathbf{x}(t)^T \mathbf{P}(t) \left(\mathbf{A} \mathbf{x}(t) - \mathbf{B} \operatorname{diag}(\mathbf{v}) \mathbf{C} \mathbf{x}(t) + \mathbf{g}(t) \right) \\
&\quad + \mathbf{p}(t)^T \left(\mathbf{A} \mathbf{x}(t) - \mathbf{B} \operatorname{diag}(\mathbf{v}) \mathbf{C} \mathbf{x}(t) + \mathbf{g}(t) \right) + \frac{1}{2} \mathbf{x}(t)^T \mathbf{Q} \mathbf{x}(t) \\
&\quad + \frac{1}{2} \sum_{i=1}^{n_u} v_i^2 r_i
\end{aligned} \tag{6.67}$$

$$\begin{aligned}
&= \frac{1}{2} \mathbf{x}(t)^T \left(\dot{\mathbf{P}}(t) + \mathbf{P}(t) \left(\mathbf{A} - \mathbf{B} \operatorname{diag}(\mathbf{v}) \mathbf{C} \right) \right. \\
&\quad \left. + \left(\mathbf{A} - \mathbf{B} \operatorname{diag}(\mathbf{v}) \mathbf{C} \right)^T \mathbf{P}(t) + \mathbf{Q} \right) \mathbf{x}(t) \\
&\quad + \mathbf{x}(t)^T \left(\dot{\mathbf{p}}(t) + \left(\mathbf{A} - \mathbf{B} \operatorname{diag}(\mathbf{v}) \mathbf{C} \right)^T \mathbf{p}(t) + \mathbf{P}(t) \mathbf{g}(t) \right) \\
&\quad + \mathbf{p}(t)^T \mathbf{g}(t) + \frac{1}{2} \sum_{i=1}^{n_u} v_i^2 r_i
\end{aligned} \tag{6.68}$$

$$\begin{aligned}
&= \frac{1}{2} \mathbf{x}(t)^T \left(\dot{\mathbf{P}}(t) + \mathbf{P}(t) \mathbf{A} + \mathbf{A}^T \mathbf{P}(t) + \mathbf{Q} \right) \mathbf{x}(t) \\
&\quad + \mathbf{x}(t)^T \left(\dot{\mathbf{p}}(t) + \mathbf{A}^T \mathbf{p}(t) + \mathbf{P}(t) \mathbf{g}(t) \right) + \mathbf{p}(t)^T \mathbf{g}(t) \\
&\quad + \left(\frac{1}{2} \sum_{i=1}^{n_u} v_i^2 r_i \right) - \mathbf{x}(t)^T \mathbf{P}(t) \mathbf{B} \operatorname{diag}(\mathbf{v}) \mathbf{C} \mathbf{x}(t) \\
&\quad - \mathbf{p}(t)^T \mathbf{B} \operatorname{diag}(\mathbf{v}) \mathbf{C} \mathbf{x}(t)
\end{aligned} \tag{6.69}$$

$$\begin{aligned}
&= \frac{1}{2} \mathbf{x}(t)^T \left(\dot{\mathbf{P}}(t) + \mathbf{P}(t) \mathbf{A} + \mathbf{A}^T \mathbf{P}(t) + \mathbf{Q} \right) \mathbf{x}(t) \\
&\quad + \mathbf{x}(t)^T \left(\dot{\mathbf{p}}(t) + \mathbf{A}^T \mathbf{p}(t) + \mathbf{P}(t) \mathbf{g}(t) \right) + \mathbf{p}(t)^T \mathbf{g}(t) \\
&\quad + \frac{1}{2} \sum_{i=1}^{n_u} \left(r_i v_i^2 - 2 \mathbf{x}(t)^T \mathbf{P}(t) \mathbf{b}_i \mathbf{c}_i \mathbf{x}(t) v_i - 2 \mathbf{p}(t)^T \mathbf{b}_i \mathbf{c}_i \mathbf{x}(t) v_i \right)
\end{aligned} \tag{6.70}$$

$$\begin{aligned}
&= \frac{1}{2} \mathbf{x}(t)^T \left(\dot{\mathbf{P}}(t) + \mathbf{P}(t) \mathbf{A} + \mathbf{A}^T \mathbf{P}(t) + \mathbf{Q} \right) \mathbf{x}(t) \\
&\quad + \mathbf{x}(t)^T \left(\dot{\mathbf{p}}(t) + \mathbf{A}^T \mathbf{p}(t) + \mathbf{P}(t) \mathbf{g}(t) \right) + \mathbf{p}(t)^T \mathbf{g}(t) \\
&\quad + \frac{1}{2} \sum_{i=1}^{n_u} \left(r_i v_i^2 - 2 r_i v_i \nu_i(t, \mathbf{x}(t)) \right)
\end{aligned} \tag{6.71}$$

where v_i was defined in the lemma. Completing the square leads to

$$s(t, \mathbf{x}(t), \mathbf{v}) = \frac{1}{2}\mathbf{x}(t)^T \left(\dot{\mathbf{P}}(t) + \mathbf{P}(t)\mathbf{A} + \mathbf{A}^T\mathbf{P}(t) + \mathbf{Q}\right)\mathbf{x}(t)$$
$$+ \mathbf{x}(t)^T \left(\dot{\mathbf{p}}(t) + \mathbf{A}^T\mathbf{p}(t) + \mathbf{P}(t)\mathbf{g}(t)\right) + \mathbf{p}(t)^T\mathbf{g}(t)$$
$$+ \frac{1}{2}\sum_{i=1}^{n_u} \left(r_i(v_i - v_i(t, \mathbf{x}(t)))^2 - r_i v_i(t, \mathbf{x}(t))^2\right)$$
$$= f_3(t, \mathbf{x}(t)) + \frac{1}{2}\sum_{i=1}^{n_u} r_i \left(v_i - v_i(t, \mathbf{x}(t))\right)^2$$

where $f_3 : \mathbb{R} \times \mathbb{R}^n \to \mathbb{R}$ is some mapping independent of v_i. It follows that a minimum of $s(t, \mathbf{x}(t), \mathbf{v})$ over $\{\mathbf{v} | \mathbf{v} \in \mathscr{U}(t, \mathbf{x})\}$ is the minimum of the quadratic sum with respect to each $\{v_i | v_i \in \mathscr{U}_i(t, \mathbf{x})\}$, independently. The minimizing and admissible $v_i \in \mathscr{U}_i(t, \mathbf{x})$ is calculated as follows:

(a) $v_i(t, \mathbf{x}(t)) \leq u_{i,min}(t, \mathbf{x}(t))$, the admissible minimum is attained at $v_i = u_{i,min}(t, \mathbf{x}(t))$.

(b) $v_i(t, \mathbf{x}(t)) \geq u_{i,max}(t, \mathbf{x}(t))$, the admissible minimum is attained at $v_i = u_{i,max}(t, \mathbf{x}(t))$.

(c) otherwise the admissible minimum is attained in $v_i = v_i(t, \mathbf{x}(t))$.

The admissible minimizing $\hat{u}_i$ that corresponds to the admissible minimizing v_i is given by Eq. (6.61).

Lemma 6.13
Let $(\mathbf{x}_k, \mathbf{u}_k)$ be a given process, and let $\mathbf{P}_k$ and $\mathbf{p}_k$ be solutions of

$$\dot{\mathbf{P}}_k(t) = -\mathbf{P}_k(t)(\mathbf{A} - \mathbf{B}\,\mathrm{diag}(\mathbf{u}_k(t))\mathbf{C}) - (\mathbf{A} - \mathbf{B}\,\mathrm{diag}(\mathbf{u}_k(t))\mathbf{C})^T\mathbf{P}_k(t) - \mathbf{Q}$$
$$\mathbf{P}_k(t_f) = \mathbf{0} \tag{6.72}$$

and

$$\dot{\mathbf{p}}_k(t) = -(\mathbf{A} - \mathbf{B}\,\mathrm{diag}(\mathbf{u}_k(t))\mathbf{C})^T\mathbf{p}_k(t) - \mathbf{P}_k(t)\mathbf{g}(t); \qquad \mathbf{p}_k(t_f) = \mathbf{0} \tag{6.73}$$

then

$$q_k(t, \mathbf{x}(t)) = \frac{1}{2}\mathbf{x}(t)^T\mathbf{P}_k(t)\mathbf{x}(t) + \mathbf{p}_k(t)^T\mathbf{x}(t)$$

is an improving function and the corresponding s_k satisfies

$$s_k(t, \mathbf{x}_k(t), \mathbf{u}_k(t)) = \max_{\boldsymbol{\xi} \in \mathscr{X}(t)} s_k(t, \boldsymbol{\xi}, \mathbf{u}_k(t)) \tag{6.74}$$

Proof 6.14 According to Eq. (6.68), $s_k(t,\mathbf{x}(t),\mathbf{u}_k(t))$ is

$$s_k(t,\mathbf{x}(t),\mathbf{u}_k(t)) = \frac{1}{2}\mathbf{x}(t)^T \Big(\dot{\mathbf{P}}_k(t) + \mathbf{P}_k(t)\,(\mathbf{A} - \mathbf{B}\operatorname{diag}(\mathbf{u}_k(t))\mathbf{C})$$

$$+ (\mathbf{A} - \mathbf{B}\operatorname{diag}(\mathbf{u}_k(t))\mathbf{C})^T \mathbf{P}_k(t) + \mathbf{Q}\Big)\mathbf{x}(t)$$

$$+ \mathbf{x}(t)^T\big(\dot{\mathbf{p}}_k(t) + (\mathbf{A} - \mathbf{B}\operatorname{diag}(\mathbf{u}_k(t))\mathbf{C})^T\,\mathbf{p}_k(t) + \mathbf{P}_k(t)\mathbf{g}(t)\big)$$

$$+ \mathbf{p}_k(t)^T\mathbf{g}(t) + \frac{1}{2}\sum_{i=1}^{n_u} u_{i,k}(t)^2 r_i$$

As $\dot{\mathbf{P}}_k(t)$ and $\dot{\mathbf{p}}_k(t)$ satisfy Eq. (6.72) and (6.73), $s_k(t,\mathbf{x}(t),\mathbf{u}_k(t))$ becomes

$$s_k(t,\mathbf{x}(t),\mathbf{u}_k(t)) = \frac{1}{2}\mathbf{x}(t)^T\mathbf{0}\mathbf{x}(t) + \mathbf{x}(t)^T\mathbf{0} + \mathbf{p}_k(t)^T\mathbf{g}(t)$$

$$= \mathbf{p}_k(t)^T\mathbf{g}(t) + \frac{1}{2}\sum_{i=1}^{n_u} u_{i,k}(t)^2 r_i \qquad (6.75)$$

Since $s_k(t,\mathbf{x}(t),\mathbf{u}_k(t)) = s_k(t,\mathbf{x}_k(t),\mathbf{u}_k(t))$, it is obvious that

$$s_k(t,\mathbf{x}(t),\mathbf{u}_k(t))) \leq s_k(t,\mathbf{x}_k(t),\mathbf{u}_k(t))$$

for all $\mathbf{x}(t)$.

Remarks.

- Equation 6.75 leads to an alternative approach for computing $J(\mathbf{x}_k,\mathbf{u}_k)$. Since $s_{kf}(\mathbf{x}(t_f)) = 0$, it is possible to write

$$J(\mathbf{x}_k,\mathbf{u}_k) = J_{eq,k}(\mathbf{x}_k,\mathbf{u}_k) = q_k(0,\mathbf{x}(0)) + \int_0^{t_f} s_k(t,\mathbf{x}_k(t),\mathbf{u}_k(t))\mathrm{d}t$$

$$= \frac{1}{2}\mathbf{x}_k(0)^T\mathbf{P}_k(0)\mathbf{x}_k(0) + \mathbf{p}_k(0)^T\mathbf{x}_k(0)$$

$$+ \int_0^{t_f}\left(\mathbf{p}_k(t)^T\mathbf{g}(t) + \frac{1}{2}\sum_{i=1}^{n_u} u_{i,k}(t)^2 r_i\right)\mathrm{d}t$$

- When $\mathbf{g} = \mathbf{0}$ it follows that $\mathbf{p}_k(t) = \mathbf{0}$. Hence, the problem is reduced to a free vibrations case, shown in [41].

Similarly to Section 6.3, putting together Section 4.3 and these two lemmas, allows to compute two sequences: $\{q_k\}$ and $\{(\mathbf{x}_k,\mathbf{u}_k)\}$ such that the second one is an improving sequence. As J is non negative, it has an infimum and $\{(\mathbf{x}_k,\mathbf{u}_k)\}$ gets arbitrarily close to an optimum.

The resulted algorithm is summarized in Fig. 6.24. Its output is an arbitrary approximation of $\mathbf{P}^*$ and $\mathbf{p}^*$, which defines the optimal control law (Eq. (6.61)). It should be noted that the use of absolute value in step (4) of the iterations stage is theoretically unnecessary. Though, due to numerical computation errors, the algorithm might lose its monotonicity when trajectories are getting closer to the optimum.

The properness of q_k is guaranteed by the same reasons that were discussed in the CBBR problem, i.e., Lemmas 6.6 and 6.7. Essentially, compared to the CBBR, here the problem is much easier as the computation of $\hat{u}_{i,k}$ does not generate discontinuities.

Let $(\mathbf{x}, \mathbf{u})$ be an admissible process, defined on the interval $[0, t_f]$. Let $\mathbf{x} = \mathbf{x}_h + \mathbf{x}_g$ where the first one is a solution to $\dot{\mathbf{x}}_h(t) = \mathbf{A}_c(t)\mathbf{x}_h(t)$ for $\mathbf{x}_h(0) = \mathbf{x}(0)$ and the second one is the solution of $\dot{\mathbf{x}}_g(t) = \mathbf{A}_c(t)\mathbf{x}_g(t) + \mathbf{g}(t)$ for $\mathbf{x}_g(0) = \mathbf{0}$. The performance index can be written as:

$$J(\mathbf{x}_h + \mathbf{x}_g, \mathbf{u}) = \frac{1}{2}\int_0^{t_f} \mathbf{x}_h(t)^T \mathbf{Q}\mathbf{x}_h(t)\mathrm{d}t + \frac{1}{2}\int_0^{t_f} \mathbf{x}_g(t)^T \mathbf{Q}\mathbf{x}_g(t)\mathrm{d}t$$

$$+ \int_0^{t_f} \mathbf{x}_h(t)^T \mathbf{Q}\mathbf{x}_g(t)\mathrm{d}t + \frac{1}{2}\int_0^{t_f} \sum_{i=1}^{n_u} u_i(t)^2 r_i \mathrm{d}t$$

The following lemma allows to relate the state cost in the above formulation with a corresponding component in J_{eq}.

Lemma 6.14
Let $\mathbf{x}_h$ and $\mathbf{x}_g$ be defined by Eqs. (6.49a) and (6.49b). Let $\mathbf{P}$ be a solution of Eq. (6.49c) for $\hat{\mathbf{Q}} = \mathbf{Q}$ and let $\mathbf{p}$ be a solution of Eq. (6.49d) for $\mathbf{P}$. Then:

(a) $\frac{1}{2}\mathbf{x}(t_1)^T \mathbf{P}(t_1)\mathbf{x}(t_1) = \frac{1}{2}\int_{t_1}^{t_f} \mathbf{x}_h(t)^T \mathbf{Q}\mathbf{x}_h(t)\mathrm{d}t$

(b) $\int_{t_1}^{t_f} \mathbf{p}(t)^T \mathbf{g}(t)\mathrm{d}t = \frac{1}{2}\int_{t_1}^{t_f} \mathbf{x}_g(t)^T \mathbf{Q}\mathbf{x}_g(t)\mathrm{d}t$

(c) $\mathbf{x}(t_1)^T \mathbf{p}(t_1) = \int_{t_1}^{t_f} \mathbf{x}_h(t)^T \mathbf{Q}\mathbf{x}_g(t)\mathrm{d}t$

Proof 6.15   The proof is a straightforward result of Lemma 6.10 with $\hat{\mathbf{Q}}(t) = \mathbf{Q}$.

Input

$\mathbf{A}, \mathbf{B} = \begin{bmatrix} \mathbf{b}_1 & \mathbf{b}_2 & \dots \end{bmatrix}, \mathbf{C} = \begin{bmatrix} \mathbf{c}_1^T & \mathbf{c}_2^T & \dots \end{bmatrix}^T, \mathbf{g}, (u_{i,min})_{i=1}^{n_u}, (u_{i,max})_{i=1}^{n_u}, \mathbf{x}(0),$
$\mathbf{Q} \geq 0, (r_i | r_i > 0)_{i=1}^{n_u}.$

Initialization

(1) Select a convergence tolerance - $\varepsilon > 0$.

(2) Solve

$$\dot{\mathbf{x}}_0(t) = (\mathbf{A} - \mathbf{B}\,\mathrm{diag}(\mathbf{u}_{min}(t, \mathbf{x}_0(t))) \mathbf{C}) \mathbf{x}_0(t) + \mathbf{g}(t); \quad \mathbf{x}(0)$$

and set $\mathbf{u}_0(t) = \mathbf{u}_{min}(t, \mathbf{x}_0(t))$. Solve

$$\dot{\mathbf{P}}_0(t) = -\mathbf{P}_0(t)(\mathbf{A} - \mathbf{B}\,\mathrm{diag}(\mathbf{u}_0(t)))\mathbf{C})$$
$$\qquad - (\mathbf{A} - \mathbf{B}\,\mathrm{diag}(\mathbf{u}_0(t)))\mathbf{C})^T \mathbf{P}_0(t) - \mathbf{Q}; \quad \mathbf{P}_0(t_f) = \mathbf{0}$$
$$\dot{\mathbf{p}}_0(t) = -(\mathbf{A} - \mathbf{B}\,\mathrm{diag}(\mathbf{u}_0(t)))\mathbf{C})^T \mathbf{p}_0(t) - \mathbf{P}_0(t)\mathbf{g}(t); \quad \mathbf{p}_0(t_f) = \mathbf{0}$$

(3) Compute $J_0(\mathbf{x}_0, \mathbf{u}_0) = \frac{1}{2} \int\limits_0^{t_f} \mathbf{x}_0(t)^T \mathbf{Q}\mathbf{x}_0(t) + \sum_{i=1}^{n_u} u_{i,0}(t)^2 r_i \mathrm{d}t$

Iterations
For $k = \{0, 1, 2, \dots\}$:

(1) Propagate to the improved process by solving

$$\dot{\mathbf{x}}_{k+1}(t) = (\mathbf{A} - \mathbf{B}\,\mathrm{diag}(\hat{\mathbf{u}}_{k+1}(t, \mathbf{x}_{k+1}(t))) \mathbf{C}) \mathbf{x}_{k+1}(t) + \mathbf{g}(t)$$

to $\mathbf{x}_{k+1}(0) = \mathbf{x}(0)$, where

$$\hat{u}_{i,k+1}(t, \mathbf{x}(t)) = \begin{cases} u_{i,min}(t, \mathbf{x}(t)) & , v_{i,k}(t, \mathbf{x}(t)) \leq u_{i,min}(t, \mathbf{x}(t)) \\ u_{i,max}(t, \mathbf{x}(t)) & , v_{i,k}(t, \mathbf{x}(t)) \geq u_{i,max}(t, \mathbf{x}(t)) \\ v_{i,k}(t, \mathbf{x}(t)) & , \text{otherwise} \end{cases}$$

$$v_{i,k}(t, \mathbf{x}(t)) \triangleq \mathbf{b}_i^T (\mathbf{P}_k(t)\mathbf{x}(t) + \mathbf{p}_k(t)) c_i \mathbf{x}(t) / r_i$$

Set $\mathbf{u}_{k+1}(t) = \hat{\mathbf{u}}_{k+1}(t, \mathbf{x}_{k+1}(t))$.

(2) Solve

$$\dot{\mathbf{P}}_{k+1}(t) = -\mathbf{P}_{k+1}(t)(\mathbf{A} - \mathbf{B}\,\mathrm{diag}(\mathbf{u}_{k+1}(t))\mathbf{C})$$
$$\qquad - (\mathbf{A} - \mathbf{B}\,\mathrm{diag}(\mathbf{u}_{k+1}(t))\mathbf{C})^T \mathbf{P}_{k+1}(t) - \mathbf{Q}$$
$$\dot{\mathbf{p}}_{k+1}(t) = -(\mathbf{A} - \mathbf{B}\,\mathrm{diag}(\mathbf{u}_{k+1}(t))\mathbf{C})^T \mathbf{p}_{k+1}(t) - \mathbf{P}_{k+1}(t)\mathbf{g}(t)$$

to $\mathbf{P}_{k+1}(t_f) = \mathbf{0}$ and $\mathbf{p}_{k+1}(t_f) = \mathbf{0}$.

Figure 6.24: CBQR—Successive improvement of control process.

---

(3) Compute

$$J(\mathbf{x}_{k+1}, \mathbf{u}_{k+1}) = \frac{1}{2} \int_0^{t_f} \mathbf{x}_{k+1}(t)^T \mathbf{Q} \mathbf{x}_{k+1}(t) + \sum_{i=1}^{n_u} u_{i,k+1}(t)^2 r_i \mathrm{d}t$$

(4) If $|J(\mathbf{x}_k, \mathbf{u}_k) - J(\mathbf{x}_{k+1}, \mathbf{u}_{k+1})| < \varepsilon$, stop iterating. Otherwise - continue.

Output
$\mathbf{P}_{k+1},\ \mathbf{p}_{k+1}.$

Figure 6.24(Cont.): CBQR—Algorithm for successive improvement of control process.

Example 6.4.1: CBQR—Free Response

In this example a free response of the 20-story model from Example 5.1.1, controlled by CBQR control law, is analyzed. Two cases are considered:

Case 1: An uncontrolled structure.

Case 2: A CBQR controlled structure with 7 semi-active dampers.

In case 2, the semi-active dampers are installed in floors 1 to 7—one device at each floor. They are numbered from 1 to 7 in an increasing order, starting at the lowest floor. The model and the dampers configuration are described in Fig. 6.25. Note that here the dampers' placement was selected by a heuristic approach that points on the lower floors as the preferred locations for dampers. However, as dampers are aimed at damping physical energy, a physical viewpoint might prefer energy effectiveness as a criterion for placing dampers. Such a heuristic approach will be described in Chapter 7.

The state weighting matrix $\mathbf{Q}$ for case 2 is chosen such that:

$$\mathbf{x}(t)^T \mathbf{Q} \mathbf{x}(t) = 5 \cdot 10^{18} \sum_{i=1}^{21} d_i(t)^2$$

where $d_i(t)$ is the inter-story drift in the i-th floor. For the given model, such weighting can be achieved by letting $\mathbf{Q} = 5 \cdot 10^{18} \mathbf{N}^T \mathbf{N}$ where $\mathbf{N} \in \mathbb{R}^{n \times n}$ is defined by $(\mathbf{N})_{i,i} = 1$ for all $1 \le i \le 21$, $(\mathbf{N})_{i+1,i} = -1$ for all $2 \le i \le 20$ and $(\mathbf{N})_{i,j} = 0$ in the other elements. The control weights are $\mathbf{R} = \mathbf{I}$. Initial state

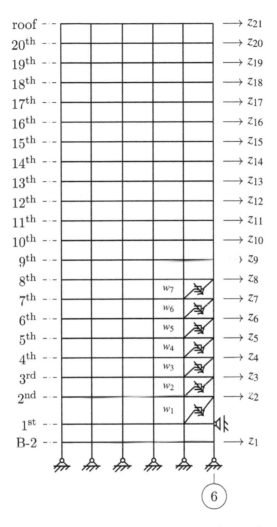

Figure 6.25: The plant's model [41, 42].

was chosen to be $z_i(0) = 0$ for all $1 \leq i \leq 21$, $\dot{z}_1(0) = 0$ and $\dot{z}_i(0) = 20 \, [\text{cm/s}]$ for all $2 \leq i \leq 21$.

The performance improvement, in iterations 1 to 17, is illustrated in Fig. 6.26. As it is evident from the figure, convergence was reached after 2 iterations and the change in J is monotonic.

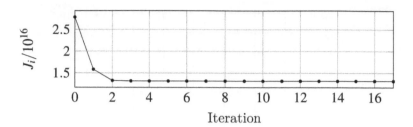

Figure 6.26: Performance index values per iteration.

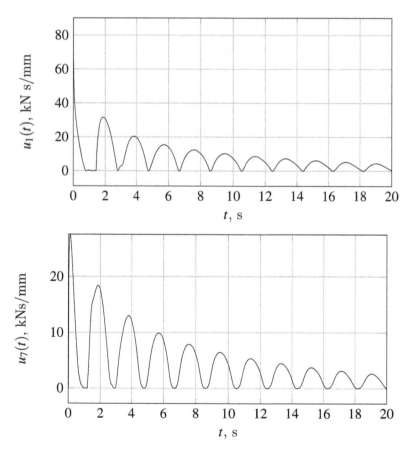

Figure 6.27: Control trajectories for case 2.

The form of control trajectories u_1 and u_7, in case 2, is illustrated in Fig. 6.27. As it can be observed, these control trajectories have a continuous form, whose practical significance should be noted. In several other semi-active control design approaches, such as bang-bang control [14, 61], the control signals are not continuous. However, there are cases, in which discontinuous signals are problematic from practical viewpoint. For example, when variable viscous dampers are used, the gains change is achieved by changing the dampers' mounting angles [78]. Discontinuous gains are not suitable in such case as these angle signals must be continuous from physical considerations.

Figure 6.28 illustrates the relation between the control force and the actuator velocity in actuators $\{1,3,5,7\}$, for case 2. It can be seen that the produced curve does not pass in the 1st and 3rd quarters of the force-velocity plane. Hence, the control power is negative. Force-elongation hysteresis plots of actuators $\{1,3,5,7\}$ are given in Fig. 6.29. It is interesting to see that the control forces vanish at peak elongations. This implies that, similar to VDs' control forces, the CBQR forces are out of the phase with the inter-story drifts.

Roof displacements are presented in Fig. 6.30. It can be seen that an improvement was achieved for case 2. Apparently, this figure shows significant similarity between case 2's response to that of a plant controlled by fluid viscous dampers. However, Fig. 6.28 demonstrates that no viscous law can be used to accurately describe the control forces. Namely, viscous dampers should generate a curved line rather than closed loops.

The peak DOFs accelerations for cases 1 and 2 are given in Fig. 6.31. The peak accelerations in case 2 were increased for DOFs 2–6, but reduced in the others. The reason for the increase is due to large control forces that were applied in these DOFs. The reduction in the other DOFs implies that the control effectively reduced the acceleration component, which is associated with the structure's motion.

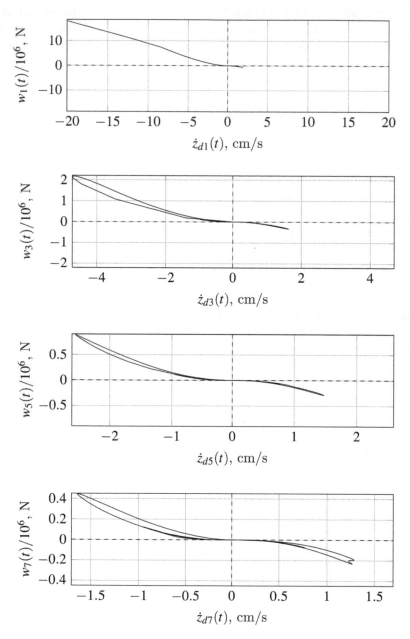

Figure 6.28: Case 2: control force vs. actuator velocity.

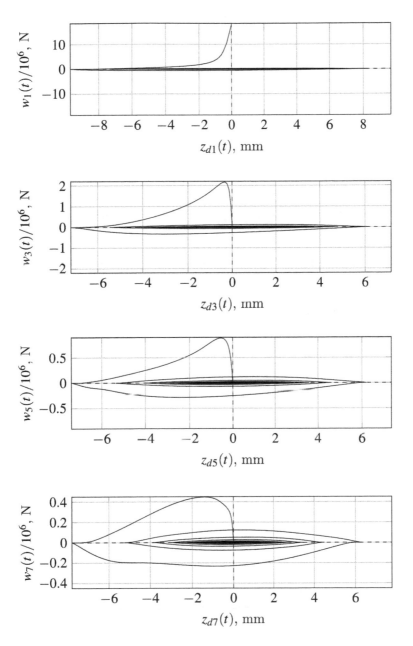

Figure 6.29: Case 2: Damper's hysteresis.

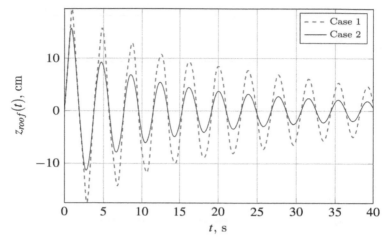

Figure 6.30: Roof displacements time history.

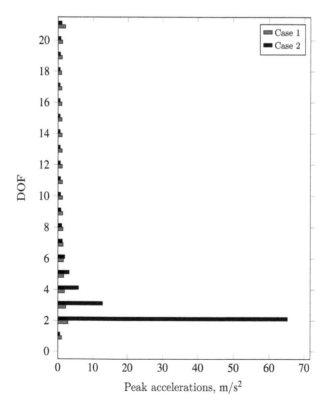

Figure 6.31: Peak DOFs accelerations.

Example 6.4.2: CBQR—Seismic Response

In this example a seismic response of the 20-story model from Example 5.1.1 is analyzed. Three cases are considered:

Case 1: An uncontrolled structure.

Case 2: A VD controlled structure.

Case 3: A CBQR controlled structure.

The response in each case is simulated to the four ground acceleration records from Example 5.2.1.

Case 2's dampers' setup is identical to that used in Example 5.1.1 for case 4. The same dampers' setup is also used here for case 3 only that semi-active dampers are considered instead of VDs. Each semi-active damper has a variable viscous gain, which is bounded below by 10^6 [kg/s] and above—by the corresponding gain's value from case 2. Additionally, the magnitude of the control force in each damper is constrained to 1.67% of the weight, carried by the floor where it is installed. Hence, for case 3:

$$u_{i,min}(t, \mathbf{x}(t)) = 1 \cdot 10^6 \text{ [kg/s]}$$

$$u_{i,max}(t, \mathbf{x}(t)) = \begin{cases} \tilde{u}_i^{c2} & , \mathbf{c}_i \mathbf{x}(t) = 0 \\ \min\left\{ \tilde{u}_i^{c2}, \frac{w_{i,max}}{\mathbf{c}_i \mathbf{x}(t)} \right\} & , \text{otherwise} \end{cases}$$

where $\tilde{u}_i^{c2}$ is the i-th gain from case 2. Table 6.3 provides the values of $\tilde{u}_i^{c2}$ and $w_{i,max}$ of each damper.

Case 3's optimal control design problem was solved by the suggested CBQR method. **Q**, the state weighting matrix, was chosen as in Example 6.4.1 and the control weighting matrix was set to $\mathbf{R} = \mathbf{I} \cdot 10^{-4}$. The initial state vector was set to zero. Four feedbacks were computed, one for each earthquake input.

The performance index's progress during the CBQR design, in each earthquake, is given in Fig. 6.32. It can be seen that convergence was reached after 3–4 iterations, depending on the earthquake input. Figures 6.33, 6.34, 6.35 and 6.36 show the roof displacements time history for each earthquake. A significant enhancement of the peak roof displacement is evident in cases 2 (a reduction of 29–38%) and 3 (a reduction of 32–40%). It can be seen that the roof displacement of cases 2 and 3 is quite similar.

Table 6.3: $\tilde{u}_i^{c2}$ and $w_{i,max}$ values for each damper.

Damper	$\tilde{u}_i^{c2}/10^6$, kg/s	Carried weight, kN	$w_{i,max}$, kN
1	1.3	56945.4	949.1
2	32.6	54340.9	905.7
3	19.9	51579.6	859.7
4	20.3	48873.1	814.6
5	19.8	46166.6	769.4
6	19.7	43460.2	724.3
7	19.7	40753.7	679.2
8	19.0	38047.3	634.1
9	18.1	35340.8	589.0
10	17.2	32634.4	543.9
11	16.2	29927.9	498.8
12	16.5	27221.5	453.7
13	16.2	24515.0	408.6
14	14.6	21808.5	363.5
15	14.0	19102.1	318.4
16	13.5	16395.6	273.3
17	11.1	13689.2	228.2
18	10.1	10982.7	183.0
19	8.8	8276.3	137.9
20	7.7	5569.8	92.8
21	6.0	2863.4	47.7

Figures 6.37, 6.38, 6.39 and 6.40 show the control trajectory, u_2, of case 3. The trajectory's form resembles that of 'bang-bang' control signals, yet it can be seen that there are moments, when it does not saturate.

Figures 6.41, 6.42, 6.43 and 6.44 presents the control force's trajectory w_2, generated in cases 2 and 3. The saturation of w_2 for case 3 can be clearly seen. Though, for case 2, the control force exceeds the allowed value.

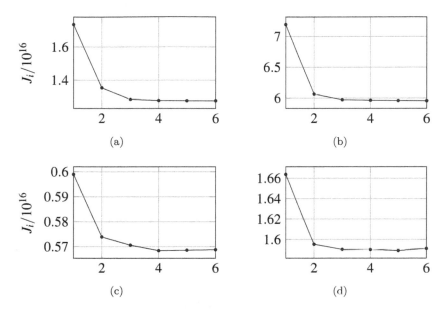

Figure 6.32: Performance index values for (a) El-Centro, (b) Hachinohe, (c) Kobe and (d) Northridge earthquakes.

Peak control forces in each earthquake are given in Figs. 6.45, 6.46, 6.47 and 6.48.   Peak inter-story drifts in each earthquake are presented in Figs. 6.49, 6.50, 6.51 and 6.52. It can be seen that better drift response was obtained in case 2, though, it involves control forces whose magnitude exceeds the upper bounds, detailed in Table 6.3.

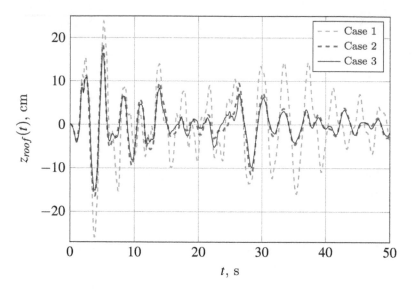

Figure 6.33: Roof displacement time history for El-Centro earthquake.

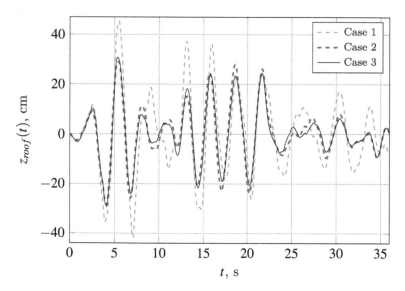

Figure 6.34: Roof displacement time history for Hachinohe earthquake.

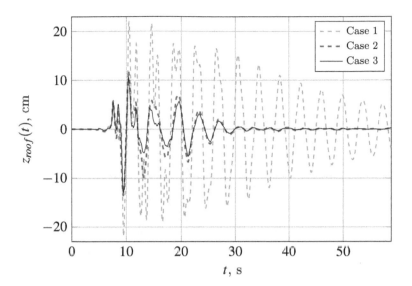

Figure 6.35: Roof displacement time history for Kobe earthquake.

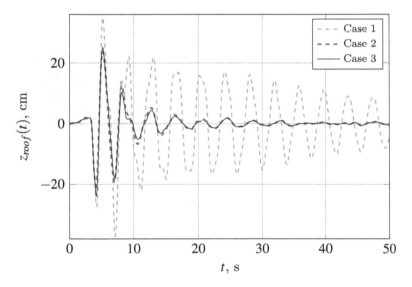

Figure 6.36: Roof displacement time history for Northridge earthquake.

Figure 6.37: Control trajectory u_2 for El-Centro earthquake.

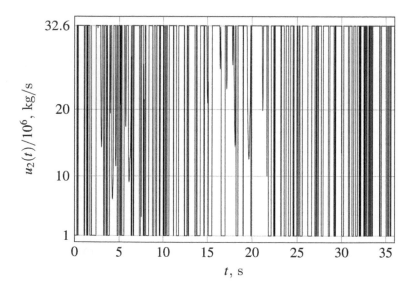

Figure 6.38: Control trajectory u_2 for Hachinohe earthquake.

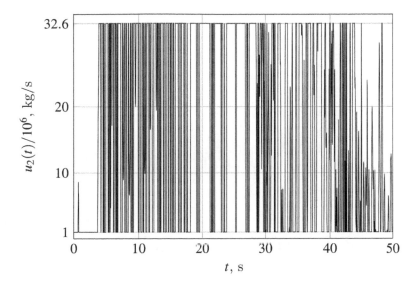

Figure 6.39: Control trajectory u_2 for Kobe earthquake.

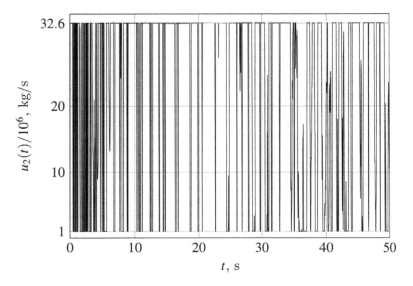

Figure 6.40: Control trajectory u_2 for Northridge earthquake.

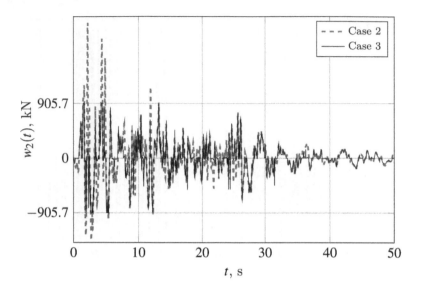

Figure 6.41: Control force trajectory w_2 for El-Centro earthquake.

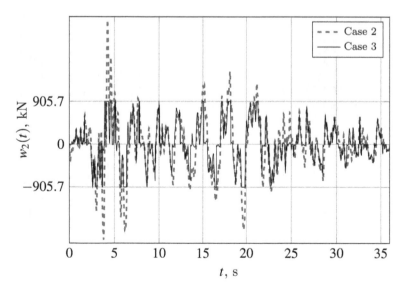

Figure 6.42: Control force trajectory w_2 for Hachinohe earthquake.

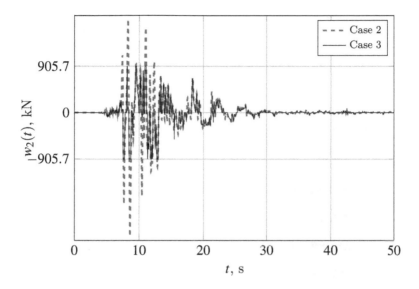

Figure 6.43: Control force trajectory w_2 for Kobe earthquake.

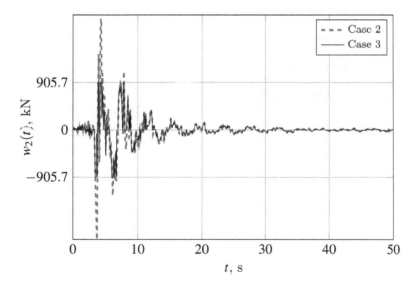

Figure 6.44: Control force trajectory w_2 for Northridge earthquake.

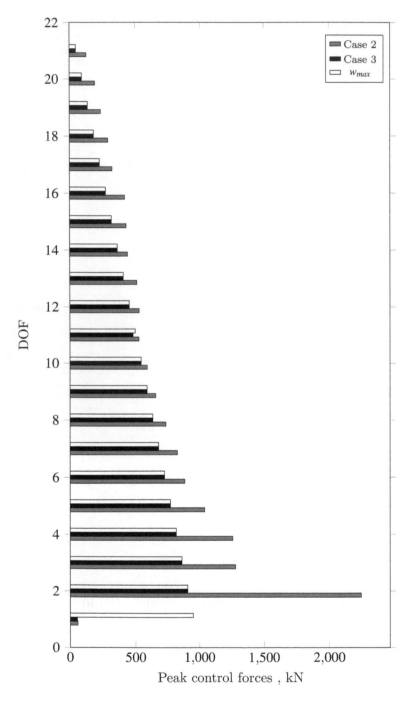

Figure 6.45: Peak control forces for El-Centro earthquake.

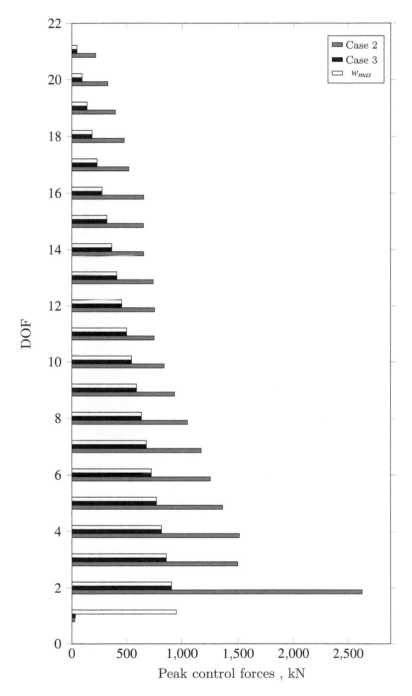

Figure 6.46: Peak control forces for Hachinohe earthquake.

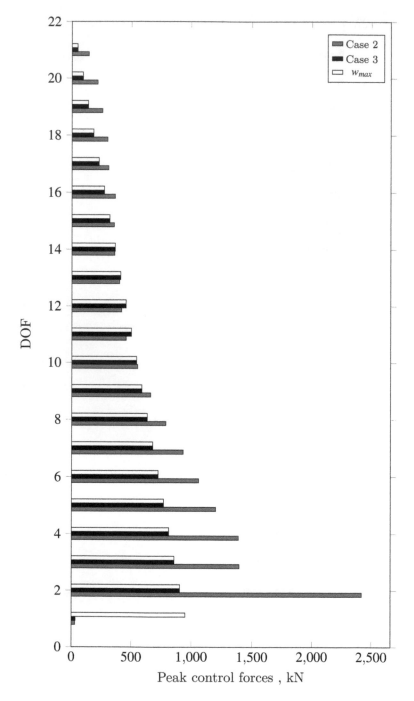

Figure 6.47: Peak control forces for Kobe earthquake.

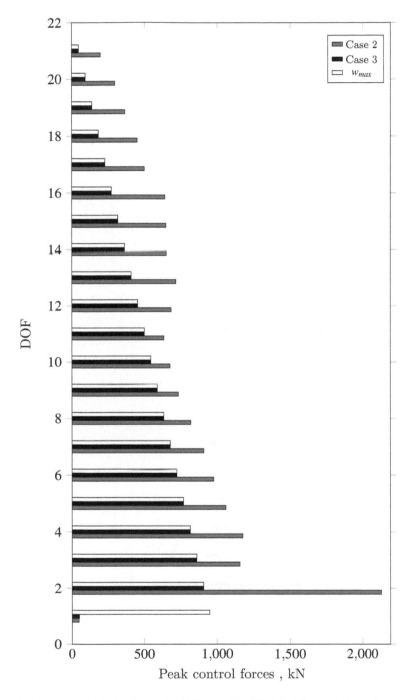

Figure 6.48: Peak control forces for Northridge earthquake.

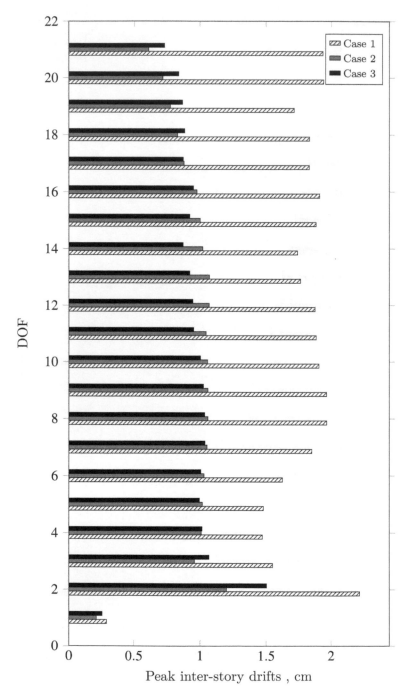

Figure 6.49: Peak inter-story drifts for El-Centro earthquake.

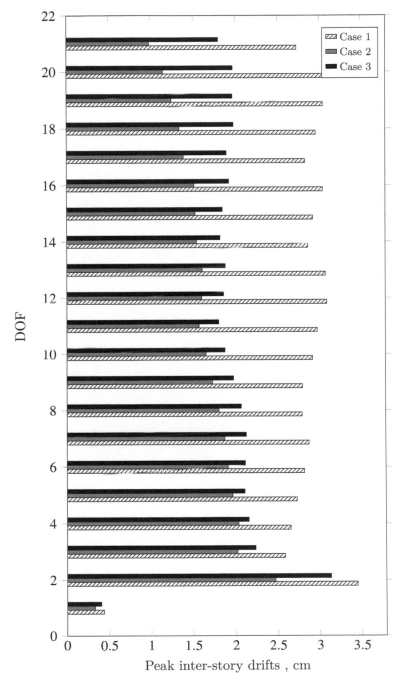

Figure 6.50: Peak inter-story drifts for Hachinohe earthquake.

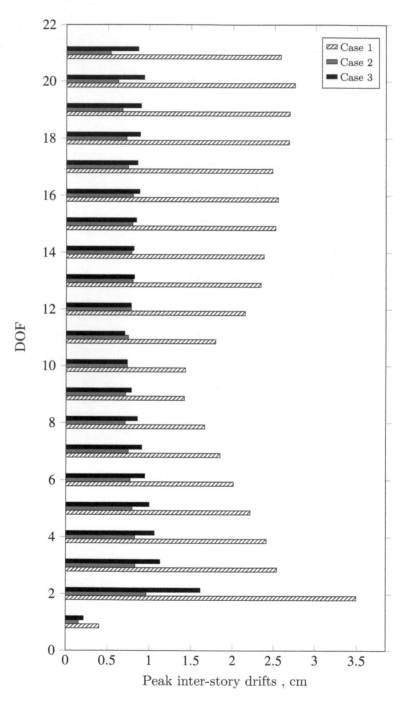

Figure 6.51: Peak inter-story drifts for Kobe earthquake.

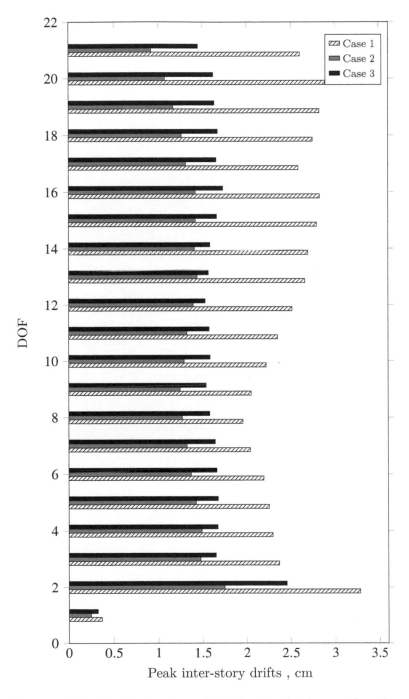

Figure 6.52: Peak inter-story drifts for Northridge earthquake.

Chapter 7

Dampers' Configuration

As previously discussed, placing dampers in structures is quite effective in improving their seismic response. However, the space required by a damper can sometimes be needed for openings, hallways, etc. Additionally, equipping a structure with a full set of dampers might have a significant impact on the overall project cost. In such cases, a structural controller with a limited set of dampers should be considered.

There may be several alternatives for placing a limited set of dampers in a given structure. Each one consists of a different selection of locations. However, even if all the locations effectively improve the structural response, their effectiveness can be different. This infers that some locations benefit more than others. Actually, it was found that for a fixed number of damping units, some configurations allow the structure to achieve its desired performance while some may not [64].

Many studies were carried out to optimize the number of dampers and their configurations, thereby improving structural seismic behavior in conjunction with cost and space saving. As a matter of fact, this is a general problem whose different aspects have been studied during the last decades [23, 104]. A research on optimal location of actuators in an actively controlled structure was conducted in the framework of the zero-one optimization problem and a constraint on the actuators number. The problem was solved by an optimization scheme, based on genetic algorithm. The maximization of energy dissipated by an active controller served as the performance index [75]. In another study it was shown that in buildings with uniform story stiffness, dampers should be placed in the lower floors [38]. A strategy for optimal selection of viscous dampers' properties was proposed, based on the system response and the control gain matrix [69]. The steepest decent method was used to find the optimal locations of viscoelastic dampers. Transfer function amplitudes of local inter-story drifts were minimized while subjected to constraint on the sum of the dampers' capacities. It was shown that the optimal configuration increases the lower mode damping ratio more effectively than uniform placement and that increasing the number of dampers unnecessarily reduces the structural response [91]. Additional solutions of this problem, in the context of seismic-resistance of structures, include: solution of an optimization problem with a statistical criterion [24]; genetic algorithms [87, 103]; and optimization by improving the dominant modes' damping ratios [64]. A method for efficiently placing active devices was suggested [2]. According to this method, the most effective locations are selected such that their contribution to the overall dissipated energy would be maximal. The method is a simple alternative to other techniques and may be easily used by the designers.

This chapter suggests a method for effectively placing semi-active dampers, using the same approach previously suggested for active devices [2]. The method points on a partial set of locations that effectively improves the seismic response compared to that obtained by applying a control system with a full set of devices. The control is designed by the CBQR optimization method, presented in the previous chapter. The configuration effectiveness is prioritized by the energy damped in it when CBQR control signals are applied.

7.1 Efficient Dampers' Configuration

The method comprises two stages—*calibration stage* and *common stage*. In both stages, the structure is designed by the CBQR method with an artificial white noise earthquake input and with dampers attached at all floors. In the calibration stage, the CBQR response is obtained while allowing all the dampers to operate. The energy dissipated by each damper is calculated and its contribution to the overall energy dissipation is computed. During the common stage, the active set of dampers is determined according to their performance in the calibration stage.

The calibration stage begins with a definition of a calibration CBQR problem. According to Eq. (2.7), a controlled structure, which is subjected to a seismic excitation, can be described by

$$\dot{\mathbf{x}}(t) = \mathbf{A}\mathbf{x}(t) + \mathbf{B}\hat{\mathbf{w}}(t,\mathbf{x}(t)) + \mathbf{g}\ddot{z}_g(t); \quad \mathbf{x}(0), \forall t \in (0,t_f)$$

where $\hat{\mathbf{w}}$ is a control force law whose output is constrained to some $\mathscr{W}(\mathbf{x})$; $\mathbf{g} \in \mathbb{R}^n$ is a ground acceleration state input vector and $\ddot{z}_g$ is an artificial white noise signal. $\ddot{z}_g$ can synthesized by the steps described in [77]. This requires to define the earthquake peak ground acceleration (PGA), the desired spectrum bandwidth (BW) and the earthquake's duration (t_f). Following the same approach, used for formulating the CBQR problem, the control force law in the i-th damper is written as $\hat{w}_i = -\hat{u}_i c_i \mathbf{x}$ where $\hat{u}_i$ is the i-th control law. Its output is constrained to a $\mathscr{U}_i(\mathbf{x})$, induced by $\mathscr{W}_i(\mathbf{x})$. In light of this, the state equation transforms to:

$$\dot{\mathbf{x}}(t) = \left(\mathbf{A} - \sum_{i=1}^{n_u} \mathbf{b}_i \hat{u}_i(t,\mathbf{x}(t)) \mathbf{c}_i \right) \mathbf{x}(t) + \mathbf{g}\ddot{z}_g(t); \quad \mathbf{x}(0), \forall t \in (0,t_f) \qquad (7.1)$$

The performance index is:

$$J(\mathbf{x},\mathbf{u}) = \frac{1}{2} \int_0^{t_f} \left(\mathbf{x}(t)^T \mathbf{Q}\mathbf{x}(t) + \sum_{i=1}^{n_u} u_i(t)^2 r_i \right) \mathrm{d}t \qquad (7.2)$$

Equations (7.1) and (7.2) define the calibration CBQR.

This calibration CBQR problem is solved by the method described in Chapter 6, thereby providing a calibration process, denoted by $(\mathbf{x}^{0*}, \mathbf{u}^{0*})$. At this point some clarification should be made regarding an ambiguity in the notation $\mathbf{g}$. The meaning of $\mathbf{g}$, used here, is different from that used in Chapter 6. In Chapter 6, $\mathbf{g}$ refers to the external state excitation trajectory (i.e., $\mathbf{g} \in \{\mathbb{R} \to \mathbb{R}^n\}$). Whereas here, it refer to a constant vector (i.e., $\mathbf{g} \in \mathbb{R}^n$), as in Eq. (2.7). I.e., it defines the distribution of a single ground acceleration input distributed between the states. It follows that the calibration CBQR should be solved with respect to the external state excitation trajectory $\mathbf{g}\ddot{z}_g$, rather then $\mathbf{g}$.

Next, the portion of energy, dissipated in each semi-active damper, is computed by integrating the damper's control force power. That is:

$$E_i \triangleq - \int_0^{t_f} w_i^{0*}(t) \mathbf{c}_i \mathbf{x}^{0*}(t) \mathrm{d}t = \int_0^{t_f} u_i^{0*}(t) (\mathbf{c}_i \mathbf{x}^{0*}(t))^2 \mathrm{d}t$$

Usually, E_i varies between damper to damper, and by that provides the main decision tool for dampers placement. The basic assumption is that locations with larger E_i are more effective for structural control. Let $\mathbf{d} = (d_1, d_2, \ldots, d_{n_u})$ be a vector of integers that describes the locations of the most effective dampers in the calibration stage, in descending order. That is:

$$E_{d_1} \geq E_{d_2} \geq E_{d_3} \geq \ldots$$

In the common stage, a common CBQR problem is solved over and over. It is similar to the calibration CBQR only that merely a partial set of dampers is allowed to operate. This set is selected according to the vector $\mathbf{d}$, as follows.

Configuration k refers to a structure with dampers installed only in locations $\mathbf{d}_{on} \triangleq (d_1, \ldots, d_k)$. The CBQR response of such structure can be computed by taking the calibration CBQR problem and disabling dampers in locations $\mathbf{d}_{off} \triangleq (d_{k+1}, \ldots, d_{n_u})$. This can be done simply by setting $\mathscr{U}_i(\mathbf{x}) \equiv 0$ for any $i \in \mathbf{d}_{off}$. The CBQR trajectory, computed with configuration k, is signified by $(\mathbf{x}^{k*}, \mathbf{u}^{k*})$. First, $(\mathbf{x}^{1*}, \mathbf{u}^{1*})$ is computed and compared to the calibration process, $(\mathbf{x}^{0*}, \mathbf{u}^{0*})$. If they are similar, the common stage terminates. Otherwise, $(\mathbf{x}^{2*}, \mathbf{u}^{2*})$ is computed and compared to $(\mathbf{x}^{0*}, \mathbf{u}^{0*})$, and so on. In order to evaluate the similarity of the processes, some external performance, J^E, should be used. That is, the common stage terminates in the smallest k that satisfies

$$\frac{|J^E(\mathbf{x}^{k*}, \mathbf{u}^{k*}) - J^E(\mathbf{x}^{0*}, \mathbf{u}^{0*})|}{J^E(\mathbf{x}^{0*}, \mathbf{u}^{0*})} < \varepsilon$$

for a prescribed $\varepsilon > 0$. The configuration that terminates the common stage, is the recommended one.

This procedure is summarized in Fig. 7.1. Its output is a vector $(d_i)_{i=1}^k$, whose elements are the preferred locations for dampers' placement.

Input

$\mathbf{A}$, $\mathbf{B} = \begin{bmatrix} \mathbf{b}_1 & \mathbf{b}_2 & \cdots & \mathbf{b}_{n_u} \end{bmatrix}$, $\mathbf{C} = \begin{bmatrix} \mathbf{c}_1^T & \mathbf{c}_2^T & \cdots & \mathbf{c}_{n_u}^T \end{bmatrix}^T$, $\mathbf{g}$, $(u_{i,min})_{i=1}^{n_u}$, $(u_{i,max})_{i=1}^{n_u}$, $\mathbf{x}(0)$, $\mathbf{Q} \geq 0$, $(r_i | r_i > 0)_{i=1}^{n_u}$, J^E, ε, PGA, BW, t_f.

Calibration Stage

(1) Generate an artificial white noise signal, $\ddot{z}_g$, according to the desired band width (BW), peak ground acceleration (PGA) and duration (t_f).

(2) Compute the calibration process $(\mathbf{x}^{0*}, \mathbf{u}^{0*})$, by solving a CBQR problem, defined by the state equation

$$\dot{\mathbf{x}}^{0*}(t) = (\mathbf{A} - \mathbf{B}\,\mathbf{diag}(\mathbf{u}^{0*}(t))\mathbf{C})\mathbf{x}^{0*}(t) + \mathbf{g}\ddot{z}_g(t); \quad \mathbf{x}(0)$$

control bounds $u_i^{0*}(t) \in [u_{i,min}(t,\mathbf{x}(t)), u_{i,max}(t,\mathbf{x}(t))]$ and performance index:

$$J(\mathbf{x}^{0*}, \mathbf{u}^{0*}) = \frac{1}{2} \int_0^{t_f} \left(\mathbf{x}^{0*}(t)^T \mathbf{Q} \mathbf{x}^{0*}(t) + \sum_{i=1}^{n_u} u_i^{0*}(t)^2 r_i \right) dt$$

(3) Compute $(E_i)_{i=1}^{n_u}$ by

$$E_i \triangleq \int_0^{t_f} u_i^{0*}(t)(\mathbf{c}_i \mathbf{x}^{0*}(t))^2 dt$$

(4) Sort $(E_i)_{i=1}^{n_u}$ and define its order by a vector $\mathbf{d} \in \mathbb{N}^{n_u}$ such that

$$E_{d_1} \geq E_{d_2} \geq E_{d_3} \geq \cdots$$

Common Stage

For $k = \{1, 2, \ldots\}$:

(1) Compute the common process $(\mathbf{x}^*, \mathbf{u}^*)$ for configuration k, by solving a CBQR problem, defined by the state equation

$$\dot{\mathbf{x}}^*(t) = (\mathbf{A} - \mathbf{B}\,\mathbf{diag}(\mathbf{u}^*(t))\mathbf{C})\mathbf{x}^*(t) + \mathbf{g}\ddot{z}_g(t); \quad \mathbf{x}(0)$$

Figure 7.1: Algorithm for efficiently placing dampers in a seismically excited structure.

control bounds

$$u_i^*(t) \begin{cases} \in [u_{i,min}(t,\mathbf{x}(t)), u_{i,max}(t,\mathbf{x}(t))], & \forall i \in [d_1, d_k] \\ \equiv 0, & \text{otherwise} \end{cases}$$

and performance index:

$$J(\mathbf{x}^*, \mathbf{u}^*) = \frac{1}{2} \int\limits_0^{t_f} \left(\mathbf{x}^*(t)^T \mathbf{Q}\mathbf{x}^*(t) + \sum_{i=1}^{n_u} u_i^*(t)^2 r_i \right) dt$$

(2) If

$$\frac{\left| J^E(\mathbf{x}^*, \mathbf{u}^*) - J^E(\mathbf{x}^{0*}, \mathbf{u}^{0*}) \right|}{J^E(\mathbf{x}^{0*}, \mathbf{u}^{0*})} < \varepsilon$$

stop iterating. Otherwise - continue.

Output
$(d_i)_{i=1}^k$.

Figure 7.1(Cont.): Algorithm for efficiently placing dampers in a seismically excited structure.

Example 7.1.1: Dampers' Configuration in a 20-Story Model

This example illustrates application of the suggested procedure for obtaining a recommended dampers' configuration to the 20-story model, described in Example 5.1.1. The lower control bounds were set to zero. No upper bounds were imposed. The initial state vector was set to zero. The weighting matrices were chosen as in Example 6.4.2. The artificial white noise ground acceleration signal was generated with bandwidth of 10 [Hz], PGA of 0.3g and $t_f = 50$ [s].

The energy, damped by each device in the calibration stage, is presented in Fig. 7.2. The dissipation pattern is commensurate with that of the standard peak inter-story drift distribution. This is not surprising as $\mathbf{Q}$'s form weighs the inter-story drifts. These are typically larger in the lower floors, thereby requires to exert larger forces on them. An exception is the basement floor. Although being among the lower floors, the energy

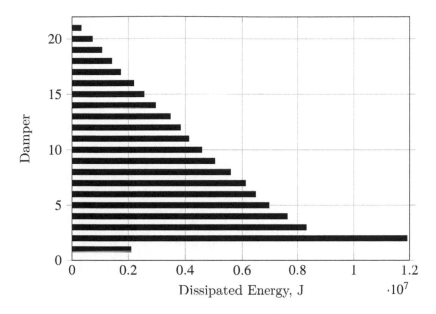

Figure 7.2: The energy damped in each device in the calibration stage.

dissipated in it is quite small. The reason for this is the horizontal support, applied by the ground floor to the basement ceiling and, by that, reducing its deformations. The dampers' placement priority, that follows the observed dissipation pattern, is:

$$\mathbf{d} = \begin{bmatrix} 2 & 3 & \dots & 16 & 1 & 17 & 18 & \dots & 21 \end{bmatrix}$$

reflecting the fact that, during the calibration stage, damper no. 1 performed better than no. 17 but worse than no. 16. This lead to 21 suggested dampers' configurations, described in Fig. 7.3. In the common stage, each dampers' configuration was embed into a suitable CBQR problem by setting the upper control bound of the inactive dampers to zero.

Each common process was assessed by an external performance index. In this case, the maximum angular inter-story drift was used for that. It is defined by:

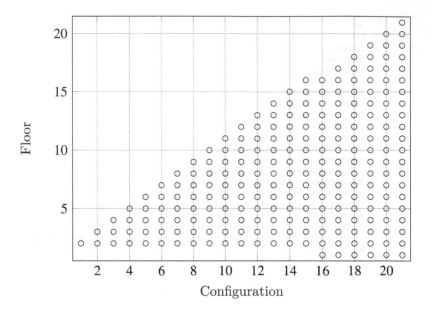

Figure 7.3: Damper configurations in a prioritized order.

$$J^E(\mathbf{z}) \triangleq \max_{\substack{t\in(0,t_f) \\ i\in[1,21]}} ISD(t,i)$$

$$ISD(t,i) \triangleq \begin{cases} \dfrac{|z_i(t)|}{h_i}, & i \in \{1,2\} \\ \dfrac{|z_i(t)-z_{i-1}(t)|}{h_i}, & \text{otherwise} \end{cases}$$

Here $\mathbf{z}$ are the displacements in the DOFs and h_i is the height of i-th floor. The performance of each common process, by means of J^E, is described in Fig. 7.4. The figure shows that, starting from the 16th process, dampers addition had merely minor contribution to the performance. In light of this, the 16th configuration was selected.

In order to portray the configuration's effectiveness, several CBQR problems were solved for three cases:

Case 1: An uncontrolled structure.

Case 2: A CBQR controlled structure with the selected configuration.

Case 3: A CBQR controlled structure with a full set of dampers.

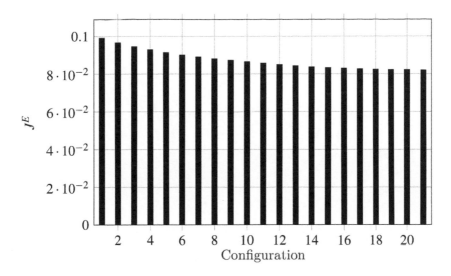

Figure 7.4: Performance of common processes obtained for the considered configurations.

Each case was designed by the CBQR method to ensemble of earthquakes: El-Centro, Hachinohe, Kobe and Northridge. The roof displacement time history, for each case and earthquake, is described in Figs. 7.5–7.8. Figures 7.9–7.12 present the peak angular inter-story drifts. It can be seen that, as far as roof displacement and the maximal peak angular inter-story drifts are concerned, the response with the selected configuration is very close to that of a structure with a full dampers' set. As a matter of fact, in large portions of the roof displacement's response, Cases 2 and 3 are indistinguishable. The selected configuration's effectiveness in reducing the response can be clearly seen.

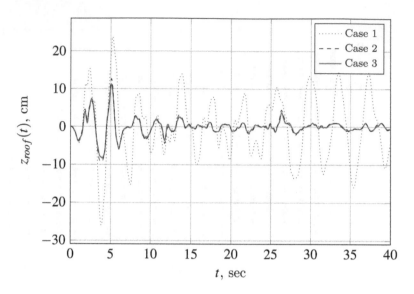

Figure 7.5: Roof displacements time history for El-Centro earthquake.

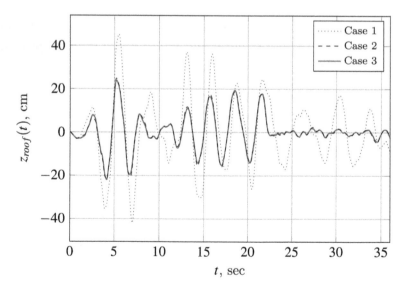

Figure 7.6: Roof displacements time history for Hachinohe earthquake.

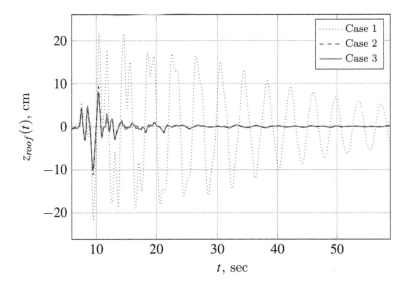

Figure 7.7: Roof displacements time history for Kobe earthquake.

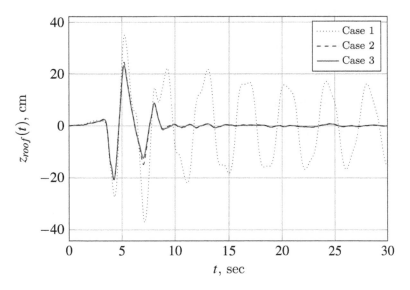

Figure 7.8: Roof displacements time history for Northridge earthquake.

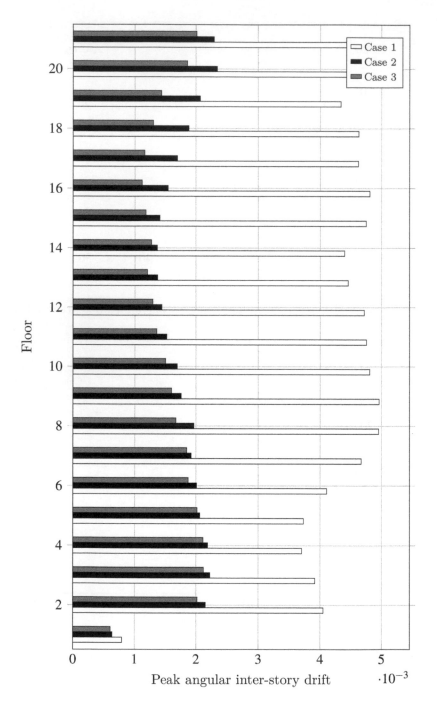

Figure 7.9: Peak angular inter-story drifts for El-Centro earthquake.

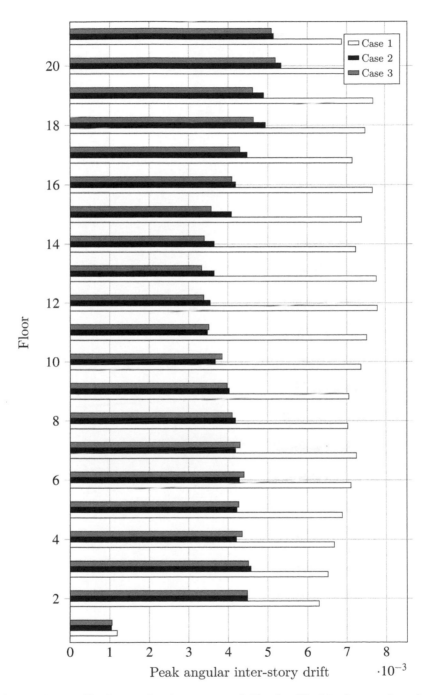

Figure 7.10: Peak angular inter-story drifts for Hachinohe earthquake.

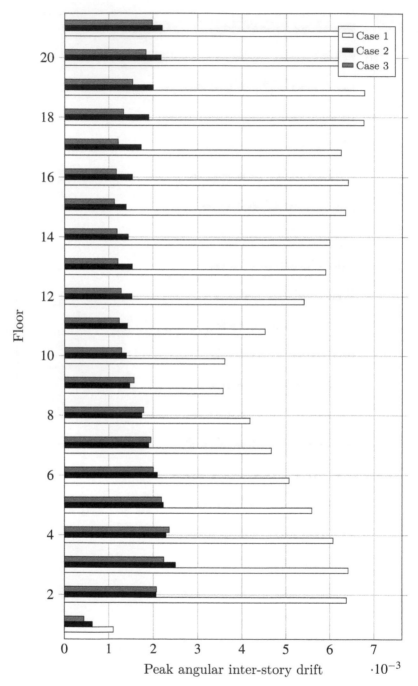

Figure 7.11: Peak angular inter-story drifts for Kobe earthquake.

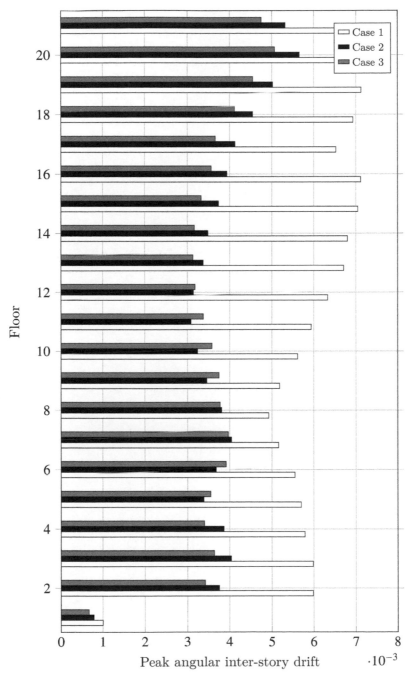

Figure 7.12: Peak angular inter-story drifts for Northridge earthquake.

Bibliography

[1] Aganovic, Z. and Z. Gajic. 1994. The successive approximation procedure for finite-time optimal control of bilinear systems. Automatic Control, IEEE Transactions on 39(9): 1932–1935.

[2] Agranovich, G. and Y. Ribakov. 2010. A method for efficient placement of active dampers in seismically excited structures. Structural Control and Health Monitoring 17(5): 513–531.

[3] Agrawal, A. and J. Yang. 2000. A semi-active electromagnetic friction damper for response control of structures. In Advanced Technology in Structural Engineering, pp. 1–8.

[4] Agrawal, A. K. and J. N. Yang. 1999. Design of passive energy dissipation systems based on lqr control methods. Journal of Intelligent Material Systems and Structures 10(12): 933–944.

[5] Aguirre, N., F. Ikhouane and J. Rodellar. 2011. Proportional-plus-integral semiactive control using magnetorheological dampers. Journal of Sound and Vibration 330(10): 2185–2200.

[6] Aliev, F. and V. Larin. 2011. Stabilization problems for a system with output feedback (review). International Applied Mechanics 47(3): 225–267.

[7] Alotta, G., L. Cavaleri, M. Di Paola and M. F. Ferrotto. 2016. Solutions for the design and increasing of efficiency of viscous dampers. The Open Construction and Building Technology Journal 10(Suppl 1:M6): 106–121.

[8] Antsaklis, P. and A. Michel. 2005. Linear Systems. Birkhäuser Boston.

[9] Bajkowski, J., J. Nachman, M. Shillor and M. Sofonea. 2008. A model for a magnetorheological damper. Mathematical and Computer Modelling 48(1-2): 56–68.

[10] Bao, J. and P. L. Lee. 2007. Process Control—The Passive Systems Approach. Springer-Verlag London.

[11] Berton, S., H. Strandgaard and J. E. Bolander. 2004. Effect of non-linear fluid viscous dampers on the size of expansion joints of multi-span prestressed concrete segmental box-girder bridges. In Proceedings of the 13nd World Conference on Earthquake Engineering, Canada.

[12] Brezas, P., M. C. Smith and W. Hoult. 2015. A clipped-optimal control algorithm for semi-active vehicle suspensions: Theory and experimental evaluation. Automatica 53(3): 188–194.

[13] Brogliato, B., R. Lozano, B. Maschke and O. Egeland. 2007. Dissipative Systems Analysis and Control. Communications and Control Engineering. Springer-Verlag London, 2 Edition.

[14] Bruni, C., G. DiPillo and G. Koch. 1974. Bilinear systems: An appealing class of "nearly linear" systems in theory and applications. IEEE Transactions on Automatic Control 19(4): 334–348.

[15] Bryson, A. E. and W. F. Denham. 1964. Optimal programming problems with inequality constraints. II—solution by steepest-ascent. AIAA Journal 2(1): 25–34.

[16] Bryson, A. E. and Y. C. Ho. 1969. Applied Optimal Control. Blaisdell, New York.

[17] Burachik, R. S., C. Y. Kaya and S. N. Majeed. 2014. A duality approach for solving control-constrained linear-quadratic optimal control problems. SIAM Journal on Control and Optimization 52(3): 1423–1456.

[18] Burl, J. B. 1999. Linear Optimal Control—H_2 and H_∞ Methods. Addison Wesley, Menlo Park, California.

[19] Buyakas, V. I. 1966. Optimal control of variable structure systems. Avtomat. i Telemekh 4: 57–68.

[20] Byrnes, C. I., A. Isidori and J. C. Willems. 1991. Passivity, feedback equivalence and the global stabilization of minimum phase nonlinear systems. IEEE Trans. Automatic Control 36(11): 1228–1240.

[21] Caughey, T. and M. O'Kelly. 1965. Classical normal modes in damped linear dynamic systems. Journal of Applied Mechanics 32(3): 583–588.

[22] Chanane, B. 1997. Bilinear quadratic optimal control: A recursive approach. Optimal Control Applications & Methods 18: 273–282.

[23] Chang, M. I. J. and T. T. Soong. 1980. Optimal controller placement in modal control of complex systems. Journal of Mathematical Analysis and Applications 75(2): 340–358.

[24] Cheng, F. Y., H. Jiang and X. Zhang. 2002. Optimal placement of dampers and actuators based on stochastic approach. Earthquake Engineering and Engineering Vibration 1(2): 237–249.

[25] Conway, J. B. 1990. A Course in Functional Analysis. Springer-Verlag New York, Berlin Heidelberg, NY, 2 Edition.

[26] Costanza, V. 2008. Finding initial costates in finite-horizon nonlinear-quadratic optimal control problems. Optimal Control Applications & Methods 29(3): 225–242.

[27] Craig, R. 1981. Structural Dynamics: An Introduction to Computer Methods. John Wiley.

[28] Datta, B. N. 2003. Numerical Methods for Linear Control Systems. Elsevier Academic Press.

[29] Davison, E. J. 1966. A method for simplifying linear dynamic systems. IEEE Transactions on Automatic Control AC-11(1): 93–101.

[30] Dower, P. M., W. M. McEneaney and M. Cantoni. 2016. A dynamic game approximation for a linear regulator problem with a log-barrier state constraint. In Proceedings of the 22nd International Symposium on Mathematical Theory of Networks and Systems.

[31] Fleming, J. F. 1990. Computer Analysis of Structural Systems. McGraw-Hill, Inc., New York, NY, USA.

[32] Forbes, J. R. and C. J. Damaren. 2010. Passive linear time-varying systems: State-space realizations, stability in feedback, and controller synthesis. In Proceedings of the 2010 American Control Conference, pp. 1097–1104, June 2010.

[33] Gajic, Z., M. Tahir and J. Qureshi. 1995. The Lyapunov Matrix Equation in System Stability and Control. Mathematics in Science and Engineering. Academic Press Limited, California, USA.

[34] Gavin, H., J. Hoagg and M. Dobossy. 2001. Optimal design of mr dampers. pp. 225–236. *In*: B. S. K. Kawashima and Y. Suzuki (eds.). U.S.-Japan

Workshop on Smart Structures for Improved Seismic Performance in Urban Regions, Seattle, WA, 8.

[35] Gawronski, W. 2004. Advanced Structural Dynamics and Active Control of Structures. Mechanical Engineering Series. Springer.

[36] Gluck, J. and Y. Ribakov. 2001. Semi-active friction system with amplifying braces for control of mdof structures. The Structural Design of Tall Buildings 10(2): 107–120.

[37] Gordanlnejad, F., A. Fuchs, U. Dogrour, S. Kanakas, Y. Liu, B. Hu and C. Evrensel. 2004. A new generation of magneto-rheological fluid dampers. Final progress report, Department of Mechanical Engineering, University of Nevada, Reno, Reno, Nevada 89557, 8.

[38] Hahn, G. D. and K. R. Sathiavageeswaran. 1992. Effects of added-damper distribution on the seismic response of buildings. Computers & Structures 43(5): 941–950.

[39] Halperin, I. and G. Agranovich. 2014. Optimal control with bilinear inequality constraints. Functional Differential Equations 21(3-4): 119–136.

[40] Halperin, I., G. Agranovich and Y. Ribakov. 2014. Efficient newton method for optimal viscous dampers design. In Electrical Electronics Engineers in Israel (IEEEI). IEEE 28th Convention of, pp. 1–4, December 2014.

[41] Halperin, I., G. Agranovich and Y. Ribakov. 2016. A method for computation of realizable optimal feedback for semi-active controlled structures. In EACS 2016—6th European Conference on Structural Control, pp. 1–11, July 2016.

[42] Halperin, I., G. Agranovich and Y. Ribakov. 2017. Using constrained bilinear quadratic regulator for the optimal semi-active control problem. Journal of Dynamic Systems, Measurement, and Control 139(11): 111011.

[43] Halperin, I., G. Agranovich and Y. Ribakov. 2017. Optimal control of a constrained bilinear dynamic system. Journal of Optimization Theory and Applications, pp. 1–15.

[44] Halperin, I., G. Agranovich and Y. Ribakov. 2017. Optimal control synthesis for the constrained bilinear biquadratic regulator problem. Optimization Letters, November 2017.

[45] Halperin, I., Y. Ribakov and G. Agranovich. 2016. Optimal viscous dampers gains for structures subjected to earthquakes. Structural Control and Health Monitoring 23(3): 458–469.

[46] Harvey, P. S., H. P. Gavin and J. T. Scruggs. 2012. Optimal performance of constrained control systems. Smart Materials and Structures 21(8): 085001.

[47] Hassan, M. F. and E. K. Boukas. 2008. Constrained linear quadratic regulator: continuous-time case. Nonlinear Dynamics and Systems Theory 8(1): 35–42.

[48] Hofer, E. P. and B. Tibken. 1988. An iterative method for the finite-time bilinear-quadratic control problem. Journal of Optimization Theory and Applications 57(3): 35–42.

[49] Housner, G., L. Bergman, T. Caughey, A. Chassiakos, R. Claus, S. Masri, R. Skelton, T. Soong, B. S. Jr. and J. Yao. 1997. Structural control: Past, present and future. Journal of Engineering Mechanics 123(9): 897–971.

[50] Ioannou, P. A. and J. Sun. 1996. Robust Adaptive Control. Prentice-Hall.

[51] Karnopp, D. 1990. Design principles for vibration control systems using semi-active dampers. Journal of Dynamic Systems Measurement and Control 112(3): 448–455.

[52] Kelley, H. J. and W. F. Denham. 1960. Gradient theory of optimal flight paths. ARS Journal 2(1): 947–953.

[53] Kirk, D. E. 1970. Optimal Control Theory—An Introduction. Prentice Hall, Englewood Cliffs, NJ.

[54] Kjeldsen, T. H. 2000. A contextualized historical analysis of the Kuhn-Tucker theorem in nonlinear programming: The impact of world war II. Historia Mathematica 27(4): 331–361.

[55] Krotov, V. F. 1988. A technique of global bounds in optimal control theory. Control and Cybernetics 17(2-3).

[56] Krotov, V. F. 1995. Global Methods in Optimal Control Theory. Chapman & Hall/CRC Pure and Applied Mathematics. CRC Press.

[57] Krotov, V. F., A. V. Bulatov and O. V. Baturina. 2011. Optimization of linear systems with controllable coefficients. Automation and Remote Control 72(6): 1199–1212.

[58] Kuhn, H. W. and A. W. Tucker. 1950. Nonlinear programming. pp. 481–492. *In*: J. Neyman (ed.). Proceedings of the Second Berkeley Symposium on Mathematical Statistics and Probability, Berkeley.

[59] Kwakernaak, H. and R. Sivan. 1972. Linear Optimal Control Systems. Wiley-Interscience Publication. Wiley Interscience.

[60] Shampine, L. F. and C. W. Gear. 1979. A user's view of solving stiff ordinary differential equations. SIAM Review 21(1): 1–17.

[61] Leavitt, J., F. Jabbari and J. E. Bobrow. 2007. Optimal performance of variable stiffness devices for structural control. Journal of Dynamic Systems Measurement and Control 129(2).

[62] Lee, S. H. and K. Lee. 2005. Bilinear systems controller design with approximation techniques. Journal of the Chungcheong Mathematical Society 18(1): 101–116.

[63] Leitmann, G. 1994. Semiactive control for vibration attenuation. Journal of Intelligent Material Systems and Structures 5(6): 841–846.

[64] Liu, W. 2003. Optimization Strategy for Damper Configurations of Buildings Based on Performance Indices. PhD thesis, State University of New York at Buffalo.

[65] Luca, S. G. and C. Pastia. 2009. Case study of variable orifice damper for seismic protection of structures. Buletinul Institutului Politehnic din Iasi. Sectia Constructii, Arhitectura 55(1): 39.

[66] Luenberger, D. G. 1984. Linear and Nonlinear Programming. Addison-Wesley, Reading, Massachusetts.

[67] Milani, B. E. A. 1979. On the computation of the optimal constant output feedback gains for large-scale linear time-invariant systems subjected to control structure constraints. In Proceedings of the 9th IFIP Conference on Optimization Techniques, Warsaw, volume 1, pp. 332–341.

[68] Morales-Beltran, M. and J. Paul. 2015. Technical note: Active and semiactive strategies to control building structures under large earthquake motion. Journal of Earthquake Engineering 19(7): 1086–1111.

[69] Gluck, N., A. M. Reinhorn, J. Gluck and R. Levy. 1996. Optimal design of supplemental dampers for control of structures. Journal of Structural Engineering 122(12): 1394–1399.

[70] Ogata, K. 2010. Modern Control Engineering. Prentice Hall, One Lake St., NJ, 5 Edition.

[71] Patten, W. N., C. C. Kuo, Q. He, L. Liu and R. L. Sack. 1994. Seismic structural control via hydraulic semi-active vibration dampers (savd). In Proceedings of the 1st World Conference on Structural Control.

[72] Polak, E. (ed.). 1997. Optimization: Algorithms and Consistent Approximations. Springer-Verlag New York.

[73] Preumont, A. 2011. Vibration Control of Active Structures: An Introduction (3rd Edition), volume 179. Springer.

[74] Rama Raju, K., M. Ansu and N. R. Iyer. 2014. A methodology of design for seismic performance enhancement of buildings using viscous fluid dampers. Structural Control and Health Monitoring 21(3): 342–355.

[75] Rao, S. S., T. S. Pan and V. B. Venkayya. 1991. Optimal placement of actuators in actively controlled structures using genetic algorithms. AIAA Journal 29(6): 942–943.

[76] Ribakov, Y. and G. Agranovich. 2011. A method for design of seismic resistant structures with viscoelastic dampers. The Structural Design of Tall and Special Buildings 20(5): 566–578.

[77] Ribakov, Y. and G. Agronovich. 2007. Design of hybrid base isolation systems with passive friction dampers. Eur. Earthquake Eng., 3: 48–56.

[78] Ribakov, Y., J. Gluck and A. M. Reinhorn. 2001. Active viscous damping system for control of mdof structures. Earthquake Engineering and Structural Dynamics 30(2): 195–212.

[79] Ribakov, Y. and J. Gluck. 1999. Active control of mdof structures with supplemental electrorheological fluid dampers. Earthquake Engineering and Structural Dynamics 28(2): 143–156.

[80] Robinson, W. D. 2012. A Pneumatic Semi-active Control Methodology for Vibration Control of Air Spring Based Suspension Systems. PhD thesis, Iowa State University.

[81] Rogers, L. C. 1970. Derivatives of eigenvalues and eigenvectors. AIAA Journal 8(5): 943–944.

[82] Dyke, M. K. S. S. J., B. F. Spencer Jr. and J. D. Carlson. 1996. Modeling and control of magnetorheological dampers for seismic response reduction. Smart Materials and Structures 5(5): 565.

[83] Sadek, F. and B. Mohraz. 1998. Semiactive control algorithms for structures with variable dampers. Journal of Engineering Mechanics 124(9): 981–990.

[84] Samali, B., K. C. S. Kwok, G. S. Wood and J. N. Yang. 2004. Wind tunnel tests for wind-excited benchmark building. Journal of Engineering Mechanics 130(4): 447–450.

[85] Scruggs, J. T. 2004. Structural Control Using Regenerative Force Actuation Networks. PhD thesis, California Institute of Technology.

[86] Silverman, L. M. and H. E. Meadows. 1967. Controllability and observability in time-variable linear systems. SIAM Journal on Control 5(1): 64–10, 02. Copyright-Copyright] © 1967 Society for Industrial and Applied Mathematics; Last updated 2012-03-02.

[87] Singh, M. P. and L. M. Moreschi. 2002. Optimal placement of dampers for passive response control. Earthquake Engineering & Structural Dynamics 31(4): 955–976.

[88] Singh, P. 1977. Seismic Behavior of Braces and Braced Steel Frames. University of Michigan, Depatment of Civil Engineering.

[89] Spencer, B. F. J. and S. Nagarajaiah. 2003. State of the art of structural control. Journal of Structural Engineering 127(7): 845–856.

[90] Spencer, Jr. B. F., R. E. Christenson and S. J. Dyke. 1998. Next generation benchmark control problems for seismically excited buildings. In Proceedings of the 2nd World Conference on Structural Control, Japan, volume 2, pp. 1335–1360.

[91] Takewaki, I. 1997. Optimal damper placement for minimum transfer functions. Earthquake Engineering & Structural Dynamics 26(11): 1113–1124.

[92] Thomas, G. B., M. D. Weir, J. R. Hass and F. R. Giordano. 2005. Thomas' Calculus Early Transcendentals (11th Edition) (Thomas Series). Addison-Wesley Longman Publishing Co., Inc., Boston, MA, USA.

[93] Truhar, N. 2004. An efficient algorithm for damper optimization for linear vibrating systems using lyapunov equation. Journal of Computational and Applied Mathematics 172(1): 169–182.

[94] Tseng, H. E. and J. K. Hedrick. 1994. Semi-active control laws-optimal and sub-optimal. Vehicle System Dynamics 23(1): 545–569.

[95] Van Overschee, P. and B. De Moor. 2011. Subspace Identification for Linear Systems: Theory Implementation Applications. Springer London, Limited.

[96] Vidyasagar, M. 1993. Nonlinear Systems Analysis. Prentice Hall, Englewood Cliffs, NJ.

[97] Viswanadham, N. and D. P. Atherton. 1975. On invariance of degree of controllability under state feedback. IEEE Trans. Automatic Control AC-20(2): 271–273.

[98] Wang, D. H. and W. H. Liao. 2005. Semiactive controllers for magnetorheological fluid dampers. Journal of Intelligent Material Systems and Structures 16(11-12): 983–993.

[99] Wang, D. H. and W. H. Liao. 2011. Magnetorheological fluid dampers: A review of parametric modelling. Smart Materials and Structures 20(2): 023001.

[100] Wang, H. 1998. Feedback stabilization of bilinear control systems. SIAM Journal on Control and Optimization 36(5): 1669–1684.

[101] Willems, J. C. 1972. Dissipative dynamical systems Part I: General theory. Archive for Rational Mechanics and Analysis 45(5): 321–351.

[102] Willems, J. C. 1972. Dissipative dynamical systems Part II: Linear systems with quadratic supply rates. Archive for Rational Mechanics and Analysis 45(5): 352–393.

[103] Wongprasert, N. and M. D. Symans. 2004. Application of a genetic algorithm for optimal damper distribution within the nonlinear seismic benchmark building. Journal of Engineering Mechanics 130(4): 401–406.

[104] Wu, Y. W., R. B. Rice and J. N. Juang. 1979. Sensor and actuator placement for large flexible space structures. In Proceedings of the Joint Automatic Control Conference 16: 230 238.

[105] Yang, J. N., A. K. Agrawal, B. Samali and W. Jong-Cheng. 2004. Benchmark problem for response control of wind-excited tall buildings. Journal of Engineering Mechanics 130(4): 437–446.

[106] Yang, J. N., S. Lin and F. Jabbari. 2003. H2-based control strategies for civil engineering structures. Journal of Structural Control 10(3-4): 205–230.

[107] Yang, J. N., S. Lin, J. H. Kim and A. K. Agrawal. 2002. Optimal design of passive energy dissipation systems based on $\|h\|_\infty$ and $\|h\|_2$ performances. Earthquake Engineering and Structural Dynamics 31(4): 921–936.

[108] Yuen, K. V., Y. Shi, J. L. Beck and H. F. Lam. 2007. Structural protection using mr dampers with clipped robust reliability-based control. Structural and Multidisciplinary Optimization 34(5): 431–443.

[109] Zames, G. 1966. On the input-output stability of time-varying nonlinear feedback systems Part I: Conditions derived using concepts of loop gain, conicity, and positivity. IEEE Transactions on Automatic Control AC-11(2): 228–238.

[110] Zareh, S. H., A. Sarrafan, A. F. Jahromi and A. A. Khayyat. 2011. Linear quadratic gaussian application and clipped optimal algorithm using for semi active vibration of passenger car. In 2011 IEEE International Conference on Mechatronics, pp. 122–127, April 2011.

[111] Zuo, L. 2002. Optimal Control with Structure Constraints and its Application to the Design of Passive Mechanical Systems. Master's thesis, Massachusetts Institute of Technology.

Index